Sebastian Dörn

Quantum Algorithms for Graph and Algebra Problems

Sebastian Dörn

Quantum Algorithms for Graph and Algebra Problems

Algorithms for Quantum Computers

VDM Verlag Dr. Müller

Imprint

Bibliographic information by the German National Library: The German National Library lists this publication at the German National Bibliography; detailed bibliographic information is available on the Internet at http://dnb.d-nb.de.

Cover image: www.purestockx.com

Publisher:
VDM Verlag Dr. Müller Aktiengesellschaft & Co. KG , Dudweiler Landstr. 125 a, 66123 Saarbrücken, Germany,
Phone +49 681 9100-698, Fax +49 681 9100-988,
Email: info@vdm-verlag.de

Zugl.: Ulm, Uni, Diss., 2007

Produced in USA and UK by:
Lightning Source Inc., La Vergne, Tennessee, USA
Lightning Source UK Ltd., Milton Keynes, UK
BookSurge LLC, 5341 Dorchester Road, Suite 16, North Charleston, SC 29418, USA

ISBN: 978-3-639-05798-0

Acknowledgments

First of all, I would like to thank my supervisor Jacobo Torán for his support and guidance during my research on this work. I am grateful to Thomas Thierauf for refereeing my thesis. His invaluable discussions and suggestions have been reflected in many places in this research. Thanks to Jens Bolte for reading the whole thesis and refereeing it. Furthermore I would like to thank my parents and grandparents for their support and encouragement.

Oberkochen, June 2008   Sebastian Dörn

Contents

1 Introduction   7

2 Preliminaries   19

   2.1 Linear Algebra 19

   2.2 Graph Theory 20

   2.3 Group Theory 24

   2.4 Quantum Computing 25

3 Methods for Quantum Algorithms   37

   3.1 Grover Search 38

   3.2 Amplitude Amplification 50

   3.3 Discrete Quantum Walk 55

4 Methods for Quantum Lower Bounds   69

   4.1 Adversary Method 70

   4.2 Polynomial Method 72

5 Fundamental Graph Problems   77

   5.1 Basic Graph Algorithms 77

5.2   Graph Connectivity   81

5.3   Minimum Spanning Tree and Shortest Paths   86

5.4   Maximum Flow   91

6   Matching Problems   97

6.1   Unweighted Matchings   99

6.2   Weighted Matchings   109

7   Graph Traversal Problems   117

7.1   Eulerian Tour   119

7.2   Optimal Postman Tour   120

7.3   Hamiltonian Circuit   124

7.4   Travelling Salesman Tour   126

7.5   Project Scheduling   130

8   Independent Set Problems   133

8.1   Maximal Independent Set   134

8.2   Maximum Independent Set   138

9   Testing Algebraic Properties   143

9.1   The Semigroup Problem   145

9.2   The Monoid Problem   149

9.3   The Group Problem   151

9.4   The Commutativity Problem   164

9.5   The Distributivity Problem   165

10 Linear Algebra Problems   **169**

    10.1 Matrix Multiplication 170

    10.2 Matrix Power 176

    10.3 Determinant and Inverse 178

Conclusion   **183**

Appendix: Overview Quantum Complexity   **187**

Bibliography   **193**

Glossary of Notation   **209**

Index   **213**

Chapter 1

Introduction

Quantum computing is an exciting new area between theoretical computer science and quantum physics. There are many reasons for exploring quantum computing. First, quantum computing is a challenge. The computation is based on quantum mechanics. All classical computers consist on classical physics, but classical physics do not describe all phenomena in nature. Therefore it is necessary to explore the potential of quantum computation as well as the power and limitation of quantum mechanics.

Quantum computing has the potential to demonstrate that for some problems it is more powerful than classical computation. An example is Shor's [Sho97] polynomial time quantum algorithm for the factorization of range intergers. On a classical computer the best known method for factorization of a number with 300 digits takes $5 \cdot 10^{24}$ steps or with terahertz speed 150.000 years. A quantum algorithms needs only $5 \cdot 10^{10}$ steps, or with terahertz speed less than one second.

Another reason for the study of quantum computing is the fact that the miniaturization of computing continues fast and achieves the microscopic level, where the laws of the quantum world dominate. At the latest in 15 years the chip development will achieve this physical limit. The demand for computing power will grow, for example for the search on databases, weather forecasts, simulation of processes or for mathematical optimization. With quantum computers we can enlarge the power of computation and speed up many important classical algorithms.

Quantum Bits and Transformations

In quantum computing we use quantum bits for the computation. A quantum bit or shortly qubit is not necessary in the state 0 or 1 like a classical bit. The qubit exists in a superposition of these two states. This means, a general state $|\psi\rangle$ of a qubit is a vector $\alpha_0|0\rangle + \alpha_1|1\rangle$, where α_0, α_1 are two complex numbers with the property $|\alpha_0|^2 + |\alpha_1|^2 = 1$. Consequently, a qubit is a unit vector in the two dimensional space spanned by the basis states $|0\rangle$ and $|1\rangle$.

We can generalize the concept of qubits to quantum registers. A state of a quantum register of size n is the tensor product of n quantum bits, and can be written as

$$|\psi\rangle = \sum_{x \in \{0,1\}^n} \alpha_x |x\rangle \quad \text{with} \quad \sum_{x \in \{0,1\}^n} |\alpha_x|^2 = 1.$$

The size of computational state space of a quantum register is exponential in the physical size of the system.

On a quantum register two basic operations can be applied: unitary evolutions and measurement. Quantum physics requires that the evolution of quantum states are described by linear and unitary operators. A unitary operator preserves the norm, the scalar product and is invertible. To get a solution of the computational problem we have to measure the current quantum state. After a measurement of a quantum state $|\psi\rangle = \sum_x \alpha_x |x\rangle$ we get $|x\rangle$ with probability $|\alpha_x|^2$. The measurement destroys the original state, and it changes the state of the system to $|x\rangle$.

The Power of Quantum Computation

Now we answer the question, what brings its power to quantum computation. The main reason is the quantum parallelism, discovered by Deutsch [Deu85]. Quantum parallelism is a fundamental feature of all quantum algorithms. This principle allows us to evaluate a function f for distinct inputs simultaneously.

For example let $f : \{0, 1\} \rightarrow \{0, 1\}$ be a Boolean function, we show, how we can compute $f(0)$ and $f(1)$ in one step. For this task we need two qubits $|x\rangle$ and $|y\rangle$ in a quantum register, denoted by $|x, y\rangle$. With a logical gate we can transform this state with a map U_f in

$$|x, y\rangle \xrightarrow{U_f} |x, y \oplus f(x)\rangle ,$$

where $\oplus$ represents the addition modulo 2. Suppose we have a classical circuit for computing f, then there is a quantum circuit to compute U_f efficiently with a quantum computer. Let $|x\rangle = \frac{|0\rangle + |1\rangle}{\sqrt{2}}$ be a superposition state and $|y\rangle = |0\rangle$. Then it holds

$$U_f \left(\frac{|0, 0\rangle + |1, 0\rangle}{\sqrt{2}} \right) = \frac{1}{\sqrt{2}} |0, f(0)\rangle + \frac{1}{\sqrt{2}} |1, f(1)\rangle .$$

9

From this quantum state follows that we have evaluated $f(x)$ for the input $x = 0$ and $x = 1$ in one step. This feature is called the quantum parallelism. We can generalize this result. Suppose we have a quantum state of n qubits. This state can exists in a superposition of 2^n basic states, and the evaluation of $f(x)$ for all 2^n values can be performed in one step. Unfortunately, the quantum parallelism is not useful in this form, because after the measurement of the quantum state we get only one value of $f(x)$. The goal is the construction of quantum algorithms with the ability to extract information about more than one value of the superposition state.

The second reason for the power of quantum computation are entangled states. Quantum entanglement is a quantum mechanical phenomenon, in which the quantum states of two or more qubits are correlated. For example, it is possible to prepare two qubits in a single quantum state, such that if one is observed in the state $|0\rangle$, the other one will always be observed in the state $|1\rangle$, and vice versa. As a result, measurements performed on one system seem to be instantaneously influencing other systems entangled with it. Quantum entanglement is the basis for quantum algorithms, quantum cryptography and has been used for experiments in quantum teleportation.

The two principles quantum parallelism and entangled states makes quantum computing more powerful than classical computation. For more details about the application of these two features, see for example the Deutsch-Jozsa quantum algorithm [DJ92].

Historical Development of Quantum Computing

Quantum computing and quantum information theory is a fast growing area in theoretical computer science and quantum physics. We give a short overview about the milestones in this research area:

1980 Paul Benioff [Ben80] was the first person who looked at the interaction between computation and quantum mechanics. He showed that reversible unitary evolution was sufficient to realize the computational power of a Turing machine.

1982 Richard Feynman [Fey82] pointed out that difficult simulations of quantum mechanics on a classical computer seem to require exponential time. He also raised the possibility of using a computer based on quantum mechanical principles to avoid this problem.

1985 David Deutsch [Deu85] defined the quantum Turing machines, a theoretical model for quantum computing. He also found the quantum parallelism principle, the main tool for quantum computation.

1992 Deutsch and Jozsa [DJ92] considered the following problem: Given a function $f : \{0, 1, \ldots, N\} \rightarrow \{0, 1\}$. Decide if f is either constant for all inputs, or else f is balanced (i.e. f is equal to 1 for exactly half of all the possible inputs, and 0 for the other half). Deutsch and Jozsa constructed a quantum algorithm which is exponentially faster than the best known classical algorithm for this problem.

1993 Bernstein and Vazirani [BV97] showed the existence of a universal quantum Turing maschine, which is capable to simulate other quantum Turing machines in polynomial time. Yao [Yao93] showed that quantum Turing machines and quantum circuits compute in polynomial time the same class of functions.

1994 Peter Shor [Sho94] developed a quantum algorithm for factorization of integers. This algorithm is exponentially faster than any known classical algorithm for factorization of integers. With Shor's algorithm, it is possible to break cryptography procedures, like the RSA algorithm.

1996 Lov Grover [Gro96] discovered a quantum search algorithm. This algorithm gives an optimal quadratic speed-up for the search of objects in an unsorted database.

1998 Since 1998 several quantum computers for small calculations have been constructed by Chung et. al. (see [CVZLL98]) and others.

2001 Aharonov, Ambainis, Kempe and Vazirani [AAKV01] introduced quantum walks on graphs, the generalization of random walks on finite graphs to the quantum world.

2001 John Watrous [Wat01] developed polynomial time quantum algorithms for computing several problems on solvable groups.

2004 Ambainis [Amb04a] presented the first quantum algorithm which uses quantum walks and goes beyond the capability of Grover search.

2004 Dürr, Heiligman, Høyer and Mhalla [DHHM04] constructed some
optimal quantum algorithms for fundamental graph problems.

Quantum Computing and Algorithms

Quantum algorithms have the potential to demonstrate that for some
problems quantum computation is more efficient than classical computation. A goal of quantum computing is to determine for which problems
quantum computers are faster than classical computers. It is still an
open issue, whether a quantum computer can solve all problems in the
complexity class **NP** in polynomial time.

Until today, there are only two basic quantum algorithmic methods
known. The first one is Shor's [Sho94] polynomial time quantum algorithm for the factorization of integers, and the second one is Grover's
search algorithm [Gro96]. Since then, we have seen some generaliza-
tions and applications of these two basic quantum techniques. The Shor
algorithm has been generalized to quantum algorithms for the hidden
subgroup problems (see e.g. [CEMM98]). Grover's search algorithm can
be used for quantum amplitude amplification [BHMT02] and quantum
random walk search [Amb04a, Sze04a, MNRS07]. The applications of
these quantum tools is a fast growing area in quantum computing.

Quantum Algorithms for Graph and Algebra Problems

In this thesis we present new quantum algorithms for graph and algebra
problems. Our quantum algorithms for these problems use a combination

of Grover search, amplitude amplification and quantum walk search. The quantum algorithms are faster than the best known classical algorithms for the corresponding problems.

We use two quantum complexity measures for our quantum algorithms included in this thesis: the query and the time complexity. The quantum query complexity of a quantum algorithm $\mathcal{A}$ is the number of quantum queries to the input made by $\mathcal{A}$, and the quantum time complexity is the number of "basic" quantum operations (measured in the circuit size of the unitary operations) made by $\mathcal{A}$. In our work we give lower and upper bounds for the quantum query complexity of important graph and algebra problems. For some of these problems we show that the complexity bounds are tight. In graph theory, we study the complexity of algorithms for matching problems on quantum computers and compare these to the best known classical algorithms. We consider different versions of matching problems, depending on whether the graph is bipartite or not and whether the graph is unweighted or weighted. We show that on a quantum computer a maximum matching in an undirected graph can be determined polynomially faster. This result improves a maximum matching quantum algorithm by Ambainis and Špalek [AS06]. Then we present the complexity of algorithms for graph traversal problems on quantum computers. More precisely, we look at eulerian tours, optimal postman tours, hamiltonian tours, travelling salesman problem and project scheduling. In particular, we prove that the quantum algorithms for the eulerian tour and the project scheduling problem are optimal in the query model. Furthermore we give quantum complexity lower and upper bounds for independent set problems in graphs.

In the algebra part, we present quantum query and time complexity bounds for group testing problems. For a set S and a binary operation on S represented as operation table, we consider the decision problem whether a groupoid, semigroup, monoid or quasigroup is a group. We also present upper and lower bounds for testing associativity, distributivity and commutativity. In particular, we give the first application of the new quantum random walk technique by Magniez, Nayak, Roland, and Santha [MNRS07]. Then we consider several quantum query complexity bounds of some important linear algebra problems. We give tight bounds for the determinant, rank, matrix inverse, and the matrix power problem. Furthermore we present an application of the quantum walk search schema for finding more than one solution of a search problem. We apply our quantum walk to matrix multiplication, thereby improving a result by Buhrman and Špalek [BŠ06].

The motivation for studying the quantum complexity of graph and algebra problems is twofold. On the one hand side, these are fundamental and basic problems which have many applications in computer science. For example, testing if a black box is a group is very useful in cryptography. On the other hand, we can analyze how powerful are our tools for the construction of lower and upper bounds for the quantum query complexity of these problem. For many problems we can find optimal quantum algorithms by a combination of Grover search, amplitude amplification and quantum walk search. But for some problems this does not seem to work. Maybe this can be a motivation for the development of new quantum techniques.

Related Work

Quantum algorithms have been presented for several problems in computer science, graph theory and algebra. The first quantum algorithms for fundamental graph problems were presented by Dürr, Heiligman, Høyer and Mhalla [DHHM04]. They studied the quantum query complexity for minimum spanning tree, graph connectivity, strong graph connectivity and single source shortest paths in the adjacency matrix and in the list model. Magniez, Santha and Szegedy [MSS05] constructed a quantum query algorithm for finding a triangle in a graph. Some polynomial time quantum algorithms are given by Ambainis and Špalek [AS06] for the maximum matching and the network flow problem.

There are two important results for algebraic problems which use quantum walk search. The first one is the work by Magniez and Nayak [MN05] in which they quantize a classical Markov chain for testing the commutativity of a black box group. In the second result Buhrman and Špalek [BŠ06] constructed a quantum algorithm for matrix multiplication and its verification.

Organisation of the Thesis

This thesis is organized as follows: In Chapter 2 we give some basic notations, definitions and facts from linear algebra, graph theory, group theory and quantum computation. In Chapter 3 we describe three important methods for the construction of quantum algorithms. We present the quantum search algorithm by Grover [Gro96], the quantum amplitude amplification [BHMT02] and the quantum walk search technique by Magniez et al [MNRS07]. These three tools are the basis for the development of our new quantum algorithms for graph and algebra problems. In Chapter 4 we present two tools for proving quantum query lower bounds. We present the quantum adversary method by Ambainis [Amb02] and the polynomial method introduced by Beals et al. [BBCMW01]. The quantum adversary tool is very useful to prove good lower bounds for many graph and algebra problems.

The part of the thesis containing the orginal results is organized in two parts. In the first part we consider the graph problems. In Chapter 5 we give a short summary of known quantum graph algorithms by [DHHM04] and [AS06]. In Chapter 6 to 8 we study the complexity of our new algorithms for matching problems, graph traversal and independent set problems on quantum computers. In the second part of our thesis we present new quantum algorithms for algebraic problems. In Chapter 9 to 10 we consider group testing problems and prove quantum complexity bounds for important problems from linear algebra.

Chapter 2

Preliminaries

This Chapter describes some basic notations, definitions and facts that are used in this thesis. It consists of four sections concerning linear algebra, graph theory, group theory, and quantum computation.

2.1 Linear Algebra

We denote with $[n]$ the set $\{1, 2, \ldots, n\}$. Let's denote by $\mathbb{F}$ an arbitary field, $\mathbb{N}$ the set of natural numbers, $\mathbb{Z}$ the set of integers, $\mathbb{R}$ the set of real numbers and $\mathbb{C}$ the set of complex numbers. An $m \times n$ matrix over the field $\mathbb{F}$ is denoted by $A = (a_{i,j}) \in \mathbb{F}^{m \times n}$. The element at position (i, j) in the matrix A is denoted by $a_{i,j}$. For a subset $R \subseteq [m]$, let $A|_{R,*}$ the $|R| \times n$ sub-matrix of A restricted to the rows from R. Analogously, for every $S \subseteq [n]$, let $A|_{*,S}$ the $m \times |S|$ sub-matrix of A restricted to the columns from S. The matrix whose elements are purely zero is called zero-matrix, which is denoted by 0. In the case when $m = n$, A is called *square matrix* of order n. We denote by $I = I_n$ the *identity* matrix of order n. The transpose of matrix A is denoted by A^T, and the transpose complex conjugate is denoted by $A^\dagger$. A square matrix that satisfy $A = A^T$ is called *symmetric*. The *spectral gap* or *eigenvalue gap* δ of A is the difference between the largest and the second largest eigenvalue of A. Let $A = (a_{i,j}) \in \mathbb{F}^{m \times n}$ and $B = (b_{i,j}) \in \mathbb{F}^{k \times l}$, the tensor product of A and B is $A \oplus B = (a_{i,j} \cdot B) \in \mathbb{F}^{mk \times nl}$.

The *characteristic function* χ_S is defined by $\chi_S(x) = 1$, if $x \in S$ and 0 otherwise. The *Kronecker function* $\delta_{x,y}$ is 1, if $x = y$, and 0 otherwise.

Let $f : \mathbb{N} \to \mathbb{N}$ be a function, we define the function classes:

$$O(f(n)) = \{g : \mathbb{N} \to \mathbb{N} \mid \exists c \in \mathbb{R}^+, n_0 \in \mathbb{N} : g(n) \leq c \cdot f(n),\ \forall n \geq n_0\}$$

$$\Omega(f(n)) = \{g : \mathbb{N} \to \mathbb{N} \mid \exists c \in \mathbb{R}^+, n_0 \in \mathbb{N} : g(n) \geq c \cdot f(n),\ \forall n \geq n_0\}.$$

We write $g \in \Theta(f(n))$ iff $g \in O(f(n))$ and $g \in \Omega(f(n))$.

2.2 Graph Theory

Let $G = (V, E)$ be an undirected graph, with $V = V(G)$ and $E = E(G)$ we denote the set of vertices and edges of G. Let $n = |V|$ be the number of vertices and $m = |E|$ the number of edges of G. We denote with $\{u, v\}$ an undirected edge between the vertices u and v in G, and called u and v *adjacent* in G. The *neighbourhood* $N_G(v)$ of a vertex v in G is the set of all vertices w of G such that v and w are adjacent in G; the cardinality of $N_G(v)$ is called the *degree* $d_G(v)$ of v. Let $U \subset V$, we write $N_G(U)$ for the set $\bigcup_{v \in U} N_G(v)$. Let $\delta(G) := \min\{d_G(v) \mid v \in V(G)\}$ be the *minimum degree* and $\Delta(G) := \max\{d_G(v) \mid v \in V(G)\}$ the *maximum degree* of G.

Let $G = (V, E)$ be a directed graph (digraph), with (u, v) we denote a directed edge in G from vertex u to vertex v; the vertex v is called *adjacent* to vertex u in G. The number of vertices adjacent to v is called the *out-degree* of v, denoted by $d_G^+(v)$. The *in-degree* of a vertex v is the number of edges directed to v, denoted by $d_G^-(v)$.

Let $S \subset V$ ($S \subset E$) be a subset of vertices (edges), we denote with G_{-S} the graph which is obtained from G by deleting all the vertices (edges) of S and the incident edges, we write G_{-a} for $G_{-\{a\}}$.

Definition 2.2.1 Let $G = (V, E)$ be an undirected graph. A *subgraph* of G is a graph G' such that $V(G') \subseteq V(G)$ and $E(G') \subseteq E(G)$. With any subset I of $V(G)$ we associate the *vertex-induced* subgraph $G[I] := (I, \{e \in E(G) \mid e \subseteq I)$. Similarly, with any subset J of $E(G)$, we associate the *edge-induced* subgraph $G[J] := (\bigcup J, J)$, where $\bigcup J = \bigcup_{j \in J} j$. A set $V' \subset V$ is called *independent*, if $G[V']$ is edgeless. A *clique* of G is a subset $V' \subset V$ such that $G[V']$ is complete.

Definition 2.2.2 A graph $G = (V, E)$ is called *simple*, if G is an undirected graph without loops (an edge that joins a single endpoint to itself) and without multiple edges (a set of edges which the same endpoints). A *walk* (*path*) between two vertices s and t in G is a sequence $(v_1, \ldots, v_k)$, where $k \geq 1$ and $v_1, \ldots, v_k$ are (distinct) vertices of G such that $s = v_1$, $t = v_k$ and $(v_i, v_{i+1}) \in E$ for $i \in [k-1]$. A *cycle* of G is a sequence $(v_1, \ldots, v_k, v_1)$ where $k \geq 3$ and $v_1, \ldots, v_k$ are distinct vertices of G such that $(v_i, v_{i+1}) \in E$ for $i \in [k - 1]$ and $(v_k, v_1) \in E$. The *length* of a walk or path $(v_1, \ldots, v_k)$ is $k - 1$, and the length of a cycle $(v_1, \ldots, v_k, v_1)$ is k. The graph G is called *connected*, if between every pair of vertices of G there is a path. A *connected component* of G is a maximal connected subgraph of G. The number of connected components of G is denoted by $c(G)$. A *tree* is a connected graph without cycle. A subgraph G' of G is a *spanning tree* of G, if G' is a tree and $V(G) = V(G')$. A *hamiltonian path (cycle)* of a graph G is a path (cycle) that contains all the vertices of G. A *hamiltonian graph* is a graph that has a hamiltonian cycle. The graph G is called *bipartite*, if the vertices of G can be partitioned into two disjoint vertex sets in such way that no edge joins two vertices in the same set.

Definition 2.2.3 Let $G_1 = (V_1, E_1)$ and $G_2 = (V_2, E_2)$ be two graphs, G_1 and G_2 called *isomorphic*, if there is a bijection $\varphi : V_1 \to V_2$ such that $(u, v) \in E_1 \Leftrightarrow (\varphi(u), \varphi(v)) \in E_2$ for all $u, v \in V_1$.

Definition 2.2.4 The *graph categorical product* $G = G_1 \times G_2$ of two graphs G_1, G_2 is defined as follows: $V(G) = V(G_1) \times V(G_2)$, and $((g_1, g_2), (g_1', g_2')) \in E(G)$ iff $(g_1, g_1') \in E(G_1)$ and $(g_2, g_2') \in E(G_2)$.

We consider the following two models for accessing information in digraphs:

- *Adjacency matrix model:* Given is the adjacency matrix $A \in \{0, 1\}^{n \times n}$ of G with $A_{i,j} = 1$ iff $(i, j) \in E$. Weighted graphs

22

are encoded by a weight matrix, where $A_{i,j}$ is the weight of edge (i, j) and for convenience we set $A_{i,j} = \infty$ if $(i, j) \notin E$.

- *Adjacency list model:* Given are the out-degrees $d_G^+(1), \ldots, d_G^+(n)$ of the vertices and for every $i \in V$ an array of its neighbours $f_i : [d_G^+(i)] \rightarrow [n]$. The value $f_i(j)$ is the j-th neighbour of i. Weighted graphs are encoded by a sequence of functions $f_i : [d_G^+(i)] \rightarrow [n] \times \mathbb{N}$, such that if $f_i(j) = (i', w)$ then there is an edge (i, i') with weight w and i' is the j-th neighbour of the vertex i.

In undirected graphs, we replace the directed edge (u, v) by an undirected edge $\{u, v\}$, and the out-degree $d_G^+(i)$ through the degree $d_G(i)$ of the every $i \in V$.

Definition 2.2.5 A *finite Markov chain* consists of:

1. A finite set $X = \{1, \ldots, n\}$, the *state space*, the elements of X are called *states*.

2. A probability vector $(p_i)_{i \in X}$, the *starting vector*.

3. A stochastic matrix $P = (p_{i,j})_{i,j \in X}$, the *transition matrix*.

The Markov chain is a random walk $x_0, x_1, x_2, \ldots$ on the states of X, such that $\text{Prob}(x_i = y \mid x_{i-1} = x) = p_{x,y}$ and $\text{Prob}(x_0 = i) = p_i$. A Markov chain is called *irreducible*, if every state is reachable from every other state (strong connected). An irreducible Markov chain is called *aperiodic*, if there is a state x and a threshold n_0 such that for every

$n > n_0$ the probability that x is reached from x after making exactly n steps is not zero (non-bipartite). A Markov chain is called *ergodic* if it is irreducible and aperiodic.

2.3   Group Theory

Definition 2.3.1 A *group* is a set G with a binary operation $\circ : G \times G \to G$ satisfying the following conditions:

1. For all $a, b, c \in G : (a \circ b) \circ c = a \circ (b \circ c)$ (*associative law*);

2. There exists $e \in G$ (*identity*) such that $e \circ a = a \circ e = a$ for all $a \in G$;

3. For each $a \in G$ there is $a^{-1} \in G$ (*inverse*) such that $a \circ a^{-1} = a^{-1} \circ a = e$.

We write ab for $a \circ b$. A group G is called *abelian*, if $ab = ba$ for all $a, b \in G$. A *subgroup* of G is a subset $H \subseteq G$, such that H is a group under the operation $\circ$ induced by G, we write $H \leq G$. A set $R \subset G$ is called generator set of G, if every element of G is representable as a product of elements of R, we write $G = \langle R \rangle$. If $a^n = e$ for some $n \in \mathbb{N}$, then a is said to have *finite order*.

Definition 2.3.2 A *groupoid* is a finite set S with a binary operation $\circ$, denoted by $(S, \circ)$. The groupoid is called a *semigroup*, if it is associative. A *monoid* is a semigroup with an identity element e, such that $a \circ e = a = e \circ a$ for all $a \in S$. A *quasigroup* is a groupoid, where all equations

$a \circ x = b$ and $x \circ a = b$ have unique solutions, and a *loop* is a quasigroup with an identity element.

2.4 Quantum Computing

We introduce the basic model of quantum computing. For more information about quantum computing, see e.g. the textbook by Nielsen and Chuang [NC03].

State Space

In quantum computing we use quantum bits for the computation. Quantum bits are elements of the two-dimensional complex Hilbert space $\mathcal{H}_2 = \mathbb{C}^2$. In a finite dimension, a Hilbert space $\mathcal{H}$ is a vector space with a scalar product. The *computational basis* is an orthonormal basis for $\mathcal{H}$, i.e. every basis vector is normalized and the scalar product between different basis vectors is zero.

Definition 2.4.1 A general state of a *quantum bit* or *qubit* is a vector

$$|\psi\rangle := \alpha_0 |0\rangle + \alpha_1 |1\rangle, \quad \alpha_0, \alpha_1 \in \mathbb{C},$$

with $|\alpha_0|^2 + |\alpha_1|^2 = 1$. The complex numbers α_0, α_1 called the *amplitudes* of the qubit. The *state space* of a qubit is a two-dimensional Hilbert space $\mathcal{H}_2 = \mathbb{C}^2$ with computational basis $\{|0\rangle, |1\rangle\}$.

A qubit can exist in a superposition of the states $|0\rangle$ and $|1\rangle$. The amplitude α_x is related to the probability of the qubit being in the state $|x\rangle$

for $x \in \{0, 1\}$. Many different physical systems can be used to realize a qubit, for example the two different polarizations of a photon or the spin of an electron. The states $|0\rangle$ and $|1\rangle$ are represented by column vectors

$$|0\rangle = \begin{pmatrix} 1 \\ 0 \end{pmatrix} \quad \text{and} \quad |1\rangle = \begin{pmatrix} 0 \\ 1 \end{pmatrix}.$$

We denote the vector $|\psi\rangle$ as *ket*-vector and $\langle\psi| = |\psi\rangle^\dagger$ as *bra*-vector. The scalar product of $|\varphi\rangle$ and $|\psi\rangle$ is denoted by $\langle\varphi|\psi\rangle$, and the norm of $|\psi\rangle$ is defined by $\| \, |\psi\rangle \, \| = \sqrt{\langle\psi|\psi\rangle}$.

Let $\mathcal{H}_X$ and $\mathcal{H}_Y$ be the Hilbert spaces spanned by the $\{|x\rangle\}_{x \in X}$ and $\{|y\rangle\}_{y \in Y}$. The *direct sum* $\mathcal{H}_X \oplus \mathcal{H}_Y$ is the Hilbert space spanned by $\{|x\rangle\}_x \cup \{|y\rangle\}_y$. The *tensor product* $\mathcal{H}_X \otimes \mathcal{H}_Y$ is the Hilbert space spanned by $\{|x\rangle \otimes |y\rangle = |x\rangle|y\rangle = |x, y\rangle\}_{x,y}$. Let $\mathcal{H}^{\otimes 1} := \mathcal{H}$ and $\mathcal{H}^{\otimes(n+1)} := \mathcal{H} \otimes \mathcal{H}^{\otimes n}$ the *tensor power* of $\mathcal{H}$.

Definition 2.4.2 A *quantum register or quantum system* of the length m is an ordered system of m qubits. The state space of such a system is the m-fold tensor product

$$\mathcal{H}_{2^m} = \underbrace{\mathcal{H}_2 \otimes \ldots \otimes \mathcal{H}_2}_{m}$$

with the computational basis states $\{|x\rangle \mid x \in \{0, 1\}^m\}$, which can also be writen as $\{|a\rangle \mid a \in \{0, 1, \ldots, 2^m - 1\}\}$. A state of a m qubit is a vector

$$|\psi\rangle := \alpha_0 |0\rangle + \alpha_1 |1\rangle + \ldots + \alpha_{2^m - 1} |2^m - 1\rangle$$

with $|\alpha_0|^2 + |\alpha_1|^2 + \ldots + |\alpha_{2^m - 1}|^2 = 1$.

A quantum system with m qubits is specified by 2^m amplitudes. For $m = 500$ this number is larger than the estimated number of atoms in the universe.

Remark 2.4.3 The states $|0\rangle, |1\rangle, \ldots, |2^m - 1\rangle$ are represented by column vectors

$$|0\rangle = \begin{pmatrix} 1 \\ 0 \\ 0 \\ \vdots \\ 0 \end{pmatrix}, \ |1\rangle = \begin{pmatrix} 0 \\ 1 \\ 0 \\ \vdots \\ 0 \end{pmatrix}, \ldots, |2^m - 1\rangle = \begin{pmatrix} 0 \\ 0 \\ 0 \\ \vdots \\ 1 \end{pmatrix}.$$

Definition 2.4.4 Let $|\psi\rangle$ be a quantum state of $\mathcal{H}_X \otimes \mathcal{H}_Y$. If $|\psi\rangle$ can be written as $|\psi_x\rangle \otimes |\psi_y\rangle$ with $|\psi_x\rangle \in \mathcal{H}_X$ and $|\psi_y\rangle \in \mathcal{H}_Y$, then $|\psi\rangle$ is called *product state*, otherwise it called *entangled*.

All computational basis states are by definition product states, but there are states which cannot be written as a direct product of two states. A famous one is the *EPR*-pair after the inventors Einstein, Podolsky, and Rosen, which is defined by $\frac{1}{\sqrt{2}}(|00\rangle + |11\rangle)$. The EPR-pair is very useful for quantum teleportation. Most of the quantum states are entangled, since the Hilbert space on n qubits has dimension 2^n, and product states can be described by using just $2n$ complex parameters.

On a quantum system two basic operations can be applied: unitary evolution and measurement.

Evolution

Quantum physics requires that the evolution of quantum states are described by linear and unitary operators.

Definition 2.4.5 An operator U is called *linear*, if it satisfies

$$U \sum_x \alpha_x |x\rangle = \sum_x \alpha_x U |x\rangle.$$

The operator U is called *unitary*, if $UU^\dagger = U^\dagger U = I$.

Remark 2.4.6 A unitary operator U preserves the norm and the scalar product, it is reversible, i.e. $U^{-1} = U^\dagger$, and it can be diagonalized with an orthonormal set of eigenvectors, where the corresponding eigenvalues are all of absolute value 1.

Remark 2.4.7 Let $\{|x\rangle\}_{x=1}^N$ be an orthonormal basis for a Hilbert space $\mathcal{H}$. The operator U can be represented in this basis by a unitary $N \times N$ matrix $U = (u_{x,y})$ with

$$U = \sum_{x,y} u_{x,y} |x\rangle \langle y|.$$

If a quantum system is in the state $|\psi\rangle$, then after applying U, the new state $U|\psi\rangle$ is determined by the matrix-vector product of U and $|\psi\rangle$.

Example 2.4.8 Some important unitary operators for qubits are

$$M_\neg := \begin{pmatrix} 0 & 1 \\ 1 & 0 \end{pmatrix} \quad \text{and} \quad H := \begin{pmatrix} \frac{1}{\sqrt{2}} & \frac{1}{\sqrt{2}} \\ \frac{1}{\sqrt{2}} & -\frac{1}{\sqrt{2}} \end{pmatrix}.$$

The matrix $M_\neg$ is called *quantum not-gate* and H is the *Hadamard* matrix.

Let U_1 and U_2 be two operators, then $U_1 \otimes U_2 : |\varphi\rangle \otimes |\psi\rangle \rightarrow U_1 |\varphi\rangle \otimes U_2 |\psi\rangle$, and $U_1^{\otimes n} := \bigotimes_{i=1}^n U_1$.

Lemma 2.4.9 *Let H be the Hadamard transformation and $|x\rangle = |x_1 \ldots x_n\rangle$, then it holds*

$$H_n := H^{\otimes n} |x\rangle = \frac{1}{\sqrt{2^n}} \sum_{z \in \{0,1\}^n} (-1)^{x \cdot z} |z\rangle \,,$$

where $x \cdot z$ is the scalar product of the two vectors.

Proof. We use the linearity of the Hadamard transformation, then

$$H^{\otimes n} |x\rangle = \bigotimes_{i=1}^n H |x_i\rangle = \frac{1}{\sqrt{2^n}} \bigotimes_{i=1}^n \sum_{z_i \in \{0,1\}} (-1)^{x_i \cdot z_i} |z_i\rangle$$

$$= \frac{1}{\sqrt{2^n}} \sum_{z \in \{0,1\}^n} (-1)^{\sum_{i=1}^n x_i \cdot z_i} |z\rangle \,.$$

$$= \frac{1}{\sqrt{2^n}} \sum_{z \in \{0,1\}^n} (-1)^{x \cdot z} |z\rangle \,.$$

$\square$

Measurement

In the last two sections we have described quantum states and what kind of operations one can apply to them. Now we present the measurement principles to get the output of the quantum computation.

The simplest case of a quantum measurement is the measurement in the computational basis. After this measurement of a quantum state $|\psi\rangle = \sum_x \alpha_x |x\rangle$ we obtain $|x\rangle$ with probability $|\alpha_x|^2$. The measurement destroys the original state and changes the state of the system to $|x\rangle$. For example, in a qubit there are an infinite number of complex amplitudes, such that in principle a qubit can represented infinity of information, but a measurement of a qubit gives only one bit 0 or 1.

We can generalize this simple measurement principle to the projective measurement. Let $\mathcal{H} = \mathcal{H}_1 \otimes \ldots \otimes \mathcal{H}_k$ be the Hilbert space of the quantum system which is split into a direct sum of orthogonal subspaces $\mathcal{H}_i$. A quantum state $|\psi\rangle \in \mathcal{H}$ can then be expressed as

$$|\psi\rangle = \sum_{i=1}^{k} \alpha_i |\psi_{\mathcal{H}_i}\rangle,$$

where $|\psi_{\mathcal{H}_i}\rangle$ is a normalized quantum state in $\mathcal{H}_i$ (the projection of $|\psi\rangle$ into H_i), and $\sum_{i=1}^{k} |\alpha_i|^2 = 1$. Observing the state $|\psi\rangle$ will cause the following:

1. One of the $\mathcal{H}_i$ will be selected with probability $|\alpha_i|^2$.

2. After the observation, the state $|\psi\rangle$ will collapse to $|\psi_{\mathcal{H}_i}\rangle$ with amplitude one.

3. The only information that we obtain, is the subspace $\mathcal{H}_i$ which was selected. All information not in the state $|\psi_{\mathcal{H}_i}\rangle$ is lost.

Remark 2.4.10 A way of representing a projective measurement are projectors. A *projector* is a linear operator P, such that $P^2 = P$ and $P^\dagger = P$. We represent the subspace $\mathcal{H}_i$ by the projector $P_i = \Pi_{\mathcal{H}_i}$ with $\sum_{i=1}^{k} P_i = I$. If a state $|\psi\rangle$ is measured by the projective measurement $\{P_i\}_{i=1}^{k}$, then the probability of obtaining the output i is $||P_i|\psi\rangle||^2 = \langle\psi|P_i|\psi\rangle$, and the quantum state collapses to $P_i|\psi\rangle/||P_i|\psi\rangle||$.

Example 2.4.11 A measurement of all qubits of an n-qubit system in the computational basis is described by the projective measurement $\{|x\rangle\langle x| \mid x \in \{0,1\}^n\}$.

Projective measurements are not the most general measurements. A generalization is the *POVM measurement*. A POVM measurement is described by a list of positive semidefinite operators. For more information about this measurement, see e.g. [NC03].

Density Operator

The quantum measurements are for some problems not useful, because they use the concept of randomness to determine the state of the quantum system after a measurement. Sometimes it is useful to consider an object that completely describes the quantum system, and which can be manipulated deterministically. This object is the density operator:

Definition 2.4.12 Let $\{p_i, |\psi_i\rangle\}_{i=1}^k$ be an ensemble of pure states of a quantum system, where p_i is the probabilities of being in state $|\psi_i\rangle$. The *density operator* for this system is defined by

$$\rho = \sum_{i=1}^k p_i |\psi_i\rangle\langle\psi_i|.$$

If $\rho = |\psi\rangle\langle\psi|$ for some $|\psi\rangle$, we call it a *pure state*, otherwise we call it a *mixed state*.

Example 2.4.13 Suppose a quantum system is in the computational basis state $|x\rangle$, then $\rho = |x\rangle\langle x|$ contains exactly one 1 on the main diag-

31

onal and it contains 0 everywhere else. If the system is in a probabilistic mixture of computational basis states described by the probability distribution p, then $\rho = \sum_x p_x |\psi_x\rangle\langle\psi_x|$ is a diagonal operator, whose diagonal represents the probability distribution.

Quantum Query Model

Many quantum algorithms are developed for the so-called oracle or query model. In the query model, the input $x_1, \ldots, x_N$ is contained in a black box or oracle and can be accessed by queries to the black box. As a query we give i as input to the black box, and the black box outputs x_i. The goal is to compute a Boolean function $f : \{0, 1\}^N \to \{0, 1\}$ on the input bits $x = (x_1, \ldots, x_N)$ minimizing the number of queries. The classical version of this model is known as decision tree.

The quantum query model was explicitly introduced by Beals et al. [BBCMW01]. In this model we pay for accessing the oracle, but unlike the classical case, we use the power of quantum parallelism to make queries in superposition. The state of the computation is represented by $|i, b, z\rangle$, where i is the query register, b is the answer register, and z is the working register.

Definition 2.4.14 A *quantum computation* with T queries is a sequence of unitary transformations

$$U_0 \to O_x \to U_1 \to O_x \to \ldots \to U_{T-1} \to O_x \to U_T,$$

where each U_j is a unitary transformation that does not depend on the input x, and O_x are query (oracle) transformations. The operator O_x can be defined as $O_x : |i, b, z\rangle \rightarrow |i, b \oplus x_i, z\rangle$. The computations consist of the following three steps:

1. Go into the initial state $|0\rangle$,

2. Apply the transformation $U_T O_x \cdots O_x U_0$,

3. Measure the final state.

The result of the computation is the rightmost bit of the state obtained by the measurement.

In the query model of computation each query adds one to the query complexity of an algorithm, but all other computations are free. The time complexity of the algorithm is usually measured in terms of the total circuit size for the unitary operations U_i. All quantum algorithms in this thesis are bounded error.

Definition 2.4.15 The *quantum query complexity* of a quantum algorithm $\mathcal{A}$ computing a function f is the number of queries to the input black box made by $\mathcal{A}$ to compute f. The *quantum time complexity* of an algorithm $\mathcal{A}$ is the number of basic quantum operations (measured in terms of the total circuit size of the unitary operations) made by $\mathcal{A}$.

Remark 2.4.16

1. The quantum query complexity of black box computation has become a great interest in quantum computing. The black box model

provides a simple and abstract framework for the construction of quantum algorithms. All quantum algorithms can be formulated in the black box model, we can determine the speed up against classical algorithm, and we can prove lower bounds for the quantum query complexity.

2. The quantum time complexity is always at least as large as the quantum query complexity, since each query takes one unit step. A lower bound for the quantum query complexity implies also the same lower bound for the quantum time complexity.

3. For most polynomial time quantum algorithms (e.g. Grover search), the time complexity is equal to the query complexity with a log factor. An exception is the hidden subgroup problem which has polynomial query complexity, yet polynomial time algorithms are known only for some instances of the problem.

Definition 2.4.17 Let $f : \{0,1\}^N \rightarrow \{0,1\}$ be a Boolean function. An algorithm computes f *exactly*, if the output equals $f(x)$ with probability 1 for all $x \in \{0,1\}^N$. It computes f with *zero-error* if it allowed to give the answer "don't know" with probability smaller than $1/2$, and if the output is 0 or 1 then this must be correct. An algorithm computes f with *bounded-error* if the output equals $f(x)$ with probability greater than $2/3$ for all $x \in \{0,1\}^N$.

Let $D(f)$, $R_0(f)$, $R_2(f)$ $(Q_E(f), Q_0(f), Q_2(f))$ be the minimum number of queries that an exact, zero-error, or bounded-error classical (quantum) algorithm have to do to compute f. Then it follows immediately:

$$N \geq D(f) \geq Q_E(f) \geq Q_0(f) \geq Q_2(f)$$
$$N \geq D(f) \geq R_0(f) \geq R_2(f) \geq Q_2(f).$$

Between the exact classical query complexity $D(f)$ and the bounded-error quantum query complexity $Q_2(f)$ holds the gap $D(f) \leq 4096Q_2(f)^6$ proved by Beals et al. [BBCMW01].

Chapter 3

Methods for Quantum Algorithms

In this Chapter we present three important tools for the construction of quantum algorithms. First we introduce the quantum search algorithm by Grover [Gro96]. This algorithm gives an optimal quadratic speed-up for the search of objects in an unsorted database. We can speed up many algorithms by a combination of Grover search with classical algorithms. Another important ingredient for the construction of quantum algorithms is the quantum amplitude amplification [BHMT02]. Suppose we have an algorithm $\mathcal{A}$ with one-sided error and small success probability of at least ε. Classically, we need $\Theta(1/\varepsilon)$ repetitions of $\mathcal{A}$ to increase its success probability from ε to a constant. In quantum computing we need only $\Theta(\sqrt{1/\varepsilon})$ repetitions of $\mathcal{A}$ to increase the success probability to a constant. The third tool for the development of our quantum algorithms is the quantum random walk search technique. Quantum walks are the quantum counterpart of Markov chains and random walks. The discrete quantum walk is a way of formulating local quantum dynamics on a graph. The walk takes discrete steps between neighbouring vertices and is a sequence of unitary transformations.

For all these methods, we give a detailed description of their derivation and application. Our quantum algorithms for graph and algebra problems use a combination of Grover search, amplitude amplification and quantum walk search.

3.1 Grover Search

An important technique for the construction of quantum algorithms is Grover's search algorithm [Gro96]. Suppose we have a set of Boolean values of size N, on a classical computer it takes $O(N)$ steps to find an entry with value one. With the Grover algorithm we find such an entry in $O(\sqrt{N})$ steps. This is a quadratic speed-up in the search of objects in an unsorted database. Combining this quantum search procedure with classical algorithms, we can speed up many algorithms.

Grover Algorithm

In this subsection we consider search problems. A *search problem* P is a subset $P \subseteq [N]$ of the search space $[N]$. With P we associate its characteristic function $f_P : [N] \rightarrow \{0, 1\}$ with

$$
f_P(x) = \begin{cases} 1, & \text{if } x \in P, \\ 0, & \text{otherwise.} \end{cases}
$$

Any $x \in P$ is called a solution to the search problem. The task is to find one or more solutions of P. Let $k = |P|$ be the number of solutions of P.

In the classical computation we need $O(N \cdot g(N))$ steps to find a solution, where $g(N)$ is the time for computing the characteristic function of P. Now we present Grover's algorithm, which finds such a solution in time $O(\sqrt{N} \cdot g(N))$.

Algorithm 1 GROVER ALGORITHM

Input: Search function $f : [N] \to \{0,1\}$ ($N = 2^n$), $k = |\{x \in [N] \mid f(x) = 1\}|$.

Output: x^* with $f(x^*) = 1$ (if there is one).

Complexity: $O(\sqrt{N/k})$ quantum queries.

1: Initialization:

$$|\psi\rangle := \frac{1}{\sqrt{N}} \sum_{x=0}^{N-1} |x\rangle$$

2: Repeat $\frac{\pi}{4}\sqrt{N/k}$ times the Grover iteration:

    1. Phase flip:

$$U_f : |x\rangle \to (-1)^{f(x)} |x\rangle.$$

    2. Diffusion:

        (a) Apply the Hadamard transformation H_n.

        (b) Perform a phase shift

$$U_0 : |x\rangle \to -(-1)^{\delta_{x,0}} |x\rangle.$$

        (c) Apply the Hadamard transformation H_n.

3: Measure the current state $|\psi\rangle$

For proving the correctness of the GROVER ALGORITHM, we use the following simple fact:

Lemma 3.1.1 *Let $N = 2^n$ and $D_N := H_n U_0 H_n$ the diffusion operator,* *then*

$$D_N = -I_N + 2|\psi\rangle\langle\psi| \quad with \quad |\psi\rangle = \frac{1}{\sqrt{N}} \sum_{x=0}^{N-1} |x\rangle.$$

Proof. Since $D_N = H_n U_0 H_n$, it holds

$$|x\rangle \xrightarrow{H_n} \frac{1}{\sqrt{N}} \sum_{y=0}^{N-1} (-1)^{x \cdot y} |y\rangle \xrightarrow{U_0} -\frac{1}{\sqrt{N}} \sum_{y=0}^{N-1} (-1)^{x \cdot y} |y\rangle + \frac{2}{\sqrt{N}} |0\rangle$$

$$\xrightarrow{H_n} -|x\rangle + 2|\psi\rangle \langle\psi|x\rangle .$$

$\square$

Theorem 3.1.2 [Gro96] *The* GROVER ALGORITHM *algorithm finds one solution of a search problem with quantum query complexity of* $O(\sqrt{N/k})$ *and success probability of at least 1/2.*

Proof. The quantum search algorithm starts in an uniform superposition state

$$|\psi\rangle = \frac{1}{\sqrt{N}} \sum_{x=0}^{N-1} |x\rangle.$$

We show that after performing $\frac{\pi}{4}\sqrt{N/k}$ application of the Grover operator $G := D_N U_f$ with $D_N = H_n U_0 H_n$ to $|\psi\rangle$, we measure with high probability a value x^* with $f(x^*) = 1$.

Our analysis is based on the fact that a Grover iteration is a rotation in a two-dimensional subspace, defined by the two states

$$|\psi_0\rangle := \frac{1}{\sqrt{N-k}} \sum_{\substack{x=0 \\ f(x)=0}}^{N-1} |x\rangle \text{ and } |\psi_1\rangle := \frac{1}{\sqrt{k}} \sum_{\substack{x=0 \\ f(x)=1}}^{N-1} |x\rangle .$$

Then the initial state of the algorithm can be written as

$$|\psi\rangle = \sqrt{\frac{N-k}{N}}|\psi_0\rangle + \sqrt{\frac{k}{N}}|\psi_1\rangle = \cos\alpha\,|\psi_0\rangle + \sin\alpha\,|\psi_1\rangle,$$

where

$$\sin\alpha := \langle\psi, \psi_1\rangle = \sqrt{\frac{k}{N}}.$$

Now we perform one step of the Grover iteration, first the phase flip step:

$$U_f(\cos\alpha|\psi_0\rangle + \sin\alpha\,|\psi_1\rangle) = \cos\alpha\,|\psi_0\rangle - \sin\alpha\,|\psi_1\rangle.$$

Second, we apply the diffusion step with the operator $D_N = -I_N + 2|\psi\rangle\langle\psi|$:

$$|\psi_0\rangle \overset{D_N}{\rightarrow} -|\psi_0\rangle + 2\cos\alpha(\cos\alpha|\psi_0\rangle + \sin\alpha|\psi_1\rangle) = \cos 2\alpha\,|\psi_0\rangle + \sin 2\alpha\,|\psi_1\rangle$$

$$|\psi_1\rangle \overset{D_N}{\rightarrow} -|\psi_0\rangle + 2\sin\alpha(\cos\alpha|\psi_0\rangle + \sin\alpha|\psi_1\rangle) = \sin 2\alpha\,|\psi_0\rangle - \cos 2\alpha\,|\psi_1\rangle.$$

Then the Grover iteration can be written in the basis $\{|\psi_0\rangle, |\psi_1\rangle\}$ as

$$G = D_N U_f = \begin{pmatrix} \cos 2\alpha & \sin 2\alpha \\ \sin 2\alpha & -\cos 2\alpha \end{pmatrix} \cdot \begin{pmatrix} 1 & 0 \\ 0 & -1 \end{pmatrix} = \begin{pmatrix} \cos 2\alpha & -\sin 2\alpha \\ \sin 2\alpha & \cos 2\alpha \end{pmatrix}.$$

From the analysis follows that the phase flip is a reflection around $|\psi_0\rangle$ in the plane spanned by the two states $|\psi_0\rangle$ and $|\psi_1\rangle$. The diffusion is a reflection around the vector $|\psi\rangle$ in the same plane. Then, one step of the Grover iteration is the composition of these two reflections, which is a rotation in the plane by the angle 2α, see Figure 3.1. If we now perform the Grover iteration t times, it holds

$$G^t|\psi\rangle = \cos((2t+1)\alpha)|\psi_0\rangle + \sin((2t+1)\alpha)|\psi_1\rangle.$$

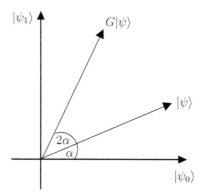

Figure 3.1: Grover iteration

Now we compute the value of t, such that the state $|\psi\rangle$ is rotated near $|\psi_1\rangle$, and then after a measurement we find a solution with high probability. For $k << N$ it is $\arcsin(\sqrt{k/N}) \approx \sqrt{k/N}$ and consequently $\alpha = \sqrt{k/N}$. Since we must take $|\psi\rangle$ near $|\psi_1\rangle$, it holds

$$(2t + 1)\alpha = \frac{\pi}{2},$$

and then it follows $t = \left\lfloor \frac{\pi}{4}\sqrt{N/k} \right\rfloor$. By a simple computation, we can show that the success probability of Grover's algorithm is at least $1/2$.

$\square$

Corollary 3.1.3 *The quantum time complexity of the* GROVER ALGORITHM *is* $O(\sqrt{N}\log N)$.

Proof. In the Grover algorithm we use $2\log n$ Hadamard transformations in each diffusion step.

$\square$

Remark 3.1.4

1. Grover [Gro02] showed that it is possible to reduce the running time $O(\log n)$ factor to $O(\log \log n)$.

2. The Grover search algorithm is optimal, i.e. there is no quantum algorithm, which can search with fewer than $\Omega(\sqrt{N})$ quantum queries (see [BBBV97] or Section 4). This gives us the information that there is no search based method for attacking **NP**-complete problems. When the **NP**-complete problems have no structure, and the best possible method for solving such a problem is a search method, then quantum computers can not solve a **NP**-complete problem efficiently.

Expected Quantum Search

In the Grover algorithm we need the number of solutions of the search problem. The problem is that we don't known this number in the most practical application. Boyer et al. [BBHT98] showed that it is possible to find a solution in $O(\sqrt{N/k})$ expected steps, even if the number k of solution is unknow.

The idea of this procedure is the following: First we start with one Grover iteration. Then we increase the number of repetitions of the iteration after every unsuccessful attempt. We choose the number of iterations from a set $[m]$. At the beginning $m = 1$, then we multiply the value of m after each choice with the factor $6/5$.

Lemma 3.1.5 [BBHT98] *Let k be the (unknown) number of solutions and let α be such that $\sin^2 \alpha = k/N$. Let m be an arbitrary positive integer and $j \in_R [0, m-1]$. After j iterations of the Grover iteration the probability of obtaining a solution is exactly*

$$P_m = \frac{1}{2} - \frac{\sin(4m\alpha)}{4m\sin(2\alpha)}.$$

In particular $P_m \geq 1/4$ when $m \geq 1/\sin(2\alpha)$.

We are now ready to describe the algorithm for finding a solution of a search problem, when k is unknown. For simplicity, we assume at first that $1 \leq k \leq 3N/4$.

Algorithm 2 EXPECTED GROVER SEARCH

Input: Search function $f : [N] \longrightarrow \{0, 1\}$.

Output: x^* with $f(x^*) = 1$.

Complexity: $O(\sqrt{N/k})$ expected quantum queries.

Comment: GROVER ALGORITHM$[f, j]$ makes j Grover iterations starting from the initial state.

  1: $m := 1$, $\lambda := 6/5$

  2: $j \in_R [0, m-1]$

  3: $x =$GROVER ALGORITHM$[f, j]$

  4: **if** $f(x) = 1$ **then**

  5:    **return**$[x]$

  6: **else**

  7:    $m := \min(\lambda m, \sqrt{N})$

  8:    **goto** 2

Theorem 3.1.6 [BBHT98] *The expected number of quantum queries of the* EXPECTED GROVER SEARCH *algorithm is $O(\sqrt{N/k})$.*

Proof. Let α be the angle such that $\sin^2 \alpha = k/N$, and

$$m_0 := \frac{1}{\sin(2\alpha)} = \frac{N}{2\sqrt{(N-k)k}} < \sqrt{\frac{N}{k}},$$

since we assumed $k \leq 3N/4$. Now we estimate the expected number of times that a Grover iteration is performed. On the i–th time round, the value of m is $\min(\lambda^{i-1}, \sqrt{N})$, and the expected number of Grover iterations is less than half this value. Note that $m < m_0$ for the first $\log_\lambda m_0$ times round the main loop. We say that the algorithm reaches the critical stage, when $m > m_0$ for the first time. The total expected number of Grover iterations needed to reach the critical stage, if it is reached, is at most

$$\frac{1}{2} \sum_{i=1}^{\lceil \log_\lambda m_0 \rceil} \lambda^{i-1} = O(m_0).$$

If the algorithm succeeds before reaching the critical stage, it finds a solution in $O(\sqrt{N/k})$ quantum queries.

Now we apply Lemma 3.1.5, if the critical stage is reached, then in every iteration from this point the search will succeed with probability of at least $1/4$. Therefore, $\frac{1}{2}\lambda^{m_0}$ expected iterations will be performed at round $m_0 + 1$. This will succeed with probability of at least $1/4$. Total the expected number of Grover iterations needed to succeed once the critical stage has been reached is

$$\frac{1}{2} \sum_{i=0}^{\infty} \left(\frac{3}{4}\right)^i \lambda^{m_0+i} = O(m_0) = O\left(\sqrt{\frac{N}{k}}\right),$$

if $0 < k \leq 3N/4$. The case $t > 3N/4$ and $t = 0$ are obviously. $\qquad \square$

Finding all Solutions

For many practical applications we are interested in all solutions of the search problem. For this task, we use the following simple procedure:

Theorem 3.1.7 *Let $P \subseteq [N]$ be a search problem and k the number of solutions of P. The quantum query complexity for finding k solutions of a search problem of size N is $O(\sqrt{N \cdot k})$.*

Proof. We use a list of all found solutions, and we modify the oracle U_f such that it flips the phase of $|x\rangle$ iff $f(x) = 1$ and x is not yet in the list. Then the query complexity for finding all solutions is

$$\sum_{i=0}^{k-1} \sqrt{\frac{N}{k-i}} = O(\sqrt{N \cdot k}).$$

$\square$

Now we summarize the results of this section in connection with quantum search bounds by Buhrman et al. [BCWZ99]:

Theorem 3.1.8 [Gro96, BBHT98, BCWZ99] *Let $P \subseteq [N]$ be a search problem and k the number of solutions of P.*

1. *Finding one solution of P can be done in $O(\sqrt{N/k})$ expected quantum queries to f_P with probability of at least a constant. The search algorithm does not require prior knowledge of k.*

2. *Finding one solution of P can be done in $O(\sqrt{N})$ quantum queries to f_P with probability of at least a constant, provided there is one.*

3. *Finding one solution of P can be done in $O(\sqrt{N \log(1/\varepsilon)})$ quantum queries to f_P with probability of at least $1 - \varepsilon$, provided there is one.*

4. *Whether $k > 0$ can be decided in $O(\sqrt{N})$ quantum queries to f_P with probability of at least a constant.*

5. *Finding all solutions of P can be done in $O(\sqrt{kN})$ quantum queries to f_P with probability of at least a constant.*

We use quantum search for speed up of classical algorithms. For these algorithms we use the following notation: Given is a search problem $P \subseteq [N]$ with $f : [N] \rightarrow \{0, 1\}$, we denote by:

- QUANTUM SEARCH$[f_P]$ an application of Grover's search algorithm, which computes an element x^* with $f_P(x^*) = 1$.

- ALL QUANTUM SEARCH$[f_P]$ an application of Grover search algorithm that computes the set $\{x \in [N] \mid f_P(x) = 1\}$ of all solutions.

Remark 3.1.9

1. The Grover search outputs a correct answer with probability of at least $1/2$. If we want to reduce the error probability to less than $1/n$, we have to repeat the quantum search $O(\log n)$ times. This increases the query complexity by a logarithmic factor. Suppose our quantum algorithm calls $l < n$ different Grover subroutines, then it outputs a correct answer with success probability at least $(1 - 1/n)^l \geq 1/e$.

2. The time complexity of our quantum algorithms which use Grover search is bigger than its query complexity by another logarithmic

factor, since the running time complexity of Grover search is logarithmic bigger than its query complexity. For simplicity, we often omit the logarithmic factors in our proofs, but we state them correctly in the statements of our theorems.

3. Suppose we have n quantum (search) algorithms, each computing some bit-value with bounded error probability. Høyer, Mosca and de Wolf [HMW03] constructed a quantum algorithm that uses $O(\sqrt{n})$ repetitions of the base algorithms and with high probability finds the index of a 1-bit among these n bits (if there is one). This shows that it is not necessary to first significantly reduce the error probability in the base algorithms to $O(1/\text{poly}(n))$.

Minima finding

Many algorithms for optimization problems utilize minima or maxima finding. We present the minima finding quantum algorithm based on quantum search by Dürr et al. [DH96, DHHM04].

Minima finding: Given an injective function $f : [N] \to \mathbb{R}$, find an index i such that $f(i)$ is a minimum of f.

For quantum search, we define a search function $f_y : [N] \to \{0, 1\}$ with

$$f_y(x) := \begin{cases} 1, & \text{if } f(x) < f(y) \\ 0, & \text{otherwise.} \end{cases}$$

The minima finding algorithm is presented in algorithm MINIMA FIND-ING. Analogously, we can compute the maxima of a function.

Algorithm 3 MINIMA FINDING

Input: Injective function $f : [N] \to \mathbb{R}$.

Output: Index i, such that $f(i)$ is a minimum of f.

Complexity: $O(\sqrt{N})$ quantum queries.

1: $j \in_R [N]$

2: **while** true **do**

3:    $i := $ QUANTUM SEARCH$[f_j]$

4:    $j := i$

5: **return**$[j]$

Theorem 3.1.10 [DH96] *The expected quantum query complexity of the* MINIMA FINDING *algorithm is* $O(\sqrt{N})$.

Proof. Every index $j \in [N]$ has a rank, which is defined as the number of indices i such that $f(i) \leq f(j)$. Suppose that the rank of j is r. Let t be a power of two such that $t \leq r \leq 2t$. After a expected number of $O(\sqrt{N/t})$ quantum queries, the algorithm finds an index i of rank less than r. The total expected number of quantum queries is at most

$$\sum_{k \geq 0}^{\log N} O\left(\sqrt{\frac{N}{2^k}} \right) = O(\sqrt{N}),$$

since $\sum_{k \geq 0} \frac{1}{\sqrt{2^k}}$ is upper bounded by a constant. The MINIMA FINDING algorithm stops, if the QUANTUM SEARCH subroutine has no solution. ☐

Next, we consider a more difficult minima finding problem:

49

Minimum type finding: Let $f, g : [N] \to \mathbb{R}$ be two functions, $e := |\{g(j) \mid j \in [N]\}|$ be the number of different types and d' be an integer. Find a subset $I \subseteq [N]$ with $d := |I| = \min\{d', e\}$ and $g(i) \neq g(i')$ for all distinct $i, i' \in I$ such that for all $j \in [N] \backslash I$ and $i \in I$, if $f(j) < f(i)$ then there exists $i' \in I$ with $f(i') \leq f(j)$ and $g(i') = g(j)$.

Theorem 3.1.11 [DHHM04] *The quantum query complexity of the minimum type finding problem is $O(\sqrt{d \cdot N})$.*

The quantum algorithm for finding the d smallest values of different types is useful for the construction of quantum algorithms for the minimum spanning tree and the shortest path problem in graphs (see Chapter 5.3).

3.2 Amplitude Amplification

The quantum amplitude amplification is a generalization of Grover's search algorithm [Gro96]. Let $\mathcal{A}$ be an algorithm with one sided error, i.e. an algorithm that if the correct answer is no, $\mathcal{A}$ always outputs no and if the correct answer is yes, $\mathcal{A}$ outputs yes with at least some probability $\varepsilon > 0$. Classically, we need $\Theta(1/\varepsilon)$ repetitions to increase its success probability from ε to a constant, for example $2/3$. With quantum computation we can perform this task quadratically better, this process is called quantum amplitude amplification.

Theorem 3.2.1 [BHMT02] *Let $\mathcal{A}$ be a quantum algorithm with one-sided error and success probability at least ε. Then there is a quantum algorithm $\mathcal{B}$ that solves $\mathcal{A}$ with success probability 2/3 by $O(\frac{1}{\sqrt{\varepsilon}})$ invocations of $\mathcal{A}$.*

Proof. Let $\mathcal{A}$ be any quantum algorithm which uses no measurements, and let

$$|\psi\rangle := \mathcal{A}|0\rangle = \sum_{x=0}^{N-1} \alpha_x |x\rangle$$

the superposition state after the application of $\mathcal{A}$. Let $S \subset [N]$ be the set of solutions of $\mathcal{A}$, and U_S a transformation that flips the phase of $|x\rangle$ if $x \in S$. We denote with $\varepsilon := \sum_{x \in S} \alpha_x$ the probability that we obtain a solution, if we measure the state in the computational basis. We show that there is a quantum algorithm $\mathcal{B}$ that uses $O(\frac{1}{\sqrt{\varepsilon}})$ applications of $\mathcal{A}$, $\mathcal{A}^{-1}, U_S$ and finds a solution with constant probability.

The amplitude amplification is similar to Grover search, but it uses the subroutines $\mathcal{A}$ and $\mathcal{A}^{-1}$ instead of the Hadamard transform. The amplification process is realized by repeatedly applying the unitary operator

$$Q := \mathcal{A} \cdot U_0 \cdot \mathcal{A}^{-1} \cdot U_S,$$

where U_0 changes the sign of the amplitude iff the state is the zero state $|0\rangle$. The operator Q is well-defined, since we assume that $\mathcal{A}$ uses no measurements, and therefore $\mathcal{A}$ has an inverse $\mathcal{A}^{-1}$.

Let $|\psi_0\rangle$ be the projection of $|\psi\rangle$ onto the bad subspace spanned by the set of basis states $|x\rangle$ for which $x \notin S$, and $|\psi_1\rangle$ is the projection of $|\psi\rangle$

onto the good subspace, spanned by the set of basis states $|x\rangle$ for which $x \in S$. We use the same arguments as in the proof of Theorem 3.1.2, the quantum state $Q^t|\psi\rangle$ stays in a two-dimensional subspace spanned by $|\psi_0\rangle$ and $|\psi_1\rangle$.

Let α_ε be defined such that $\sin^2(\alpha_\varepsilon) = \varepsilon$ and $0 < \alpha_\varepsilon \le \pi/2$. From the analysis of the Grover algorithm follows that after t applications of operator Q, the state is

$$Q^t|\psi\rangle = \frac{1}{\sqrt{\varepsilon}}\sin((2t+1)\alpha_\varepsilon)|\psi_1\rangle + \frac{1}{\sqrt{1-\varepsilon}}\cos((2t+1)\alpha_\varepsilon)|\psi_0\rangle.$$

Now we compute the value of t, such that the state $|\psi\rangle$ is rotated near $|\psi_1\rangle$, and then after a measurement we find a solution with high probability. If ε is small, then $\alpha_\varepsilon = \sqrt{\varepsilon}$. Since we must take $|\psi\rangle$ near $|\psi_1\rangle$, it holds

$$(2t+1)\alpha_\varepsilon = \frac{\pi}{2},$$

and it follows $m = \left\lfloor \frac{\pi}{4\sqrt{\varepsilon}} \right\rfloor = O(\frac{1}{\sqrt{\varepsilon}})$. $\qquad\square$

Remark 3.2.2 For the transformation of a classical algorithm into a quantum algorithm, we can use Grover search, or the fact that any classical computation can be simulated on a quantum computer (see e.g. [NC03]). More precisely, in the query model, a classical computation with N queries can be simulated by a quantum computation with N queries. The application of the amplitude amplification follows immediately:

1. Take a classical algorithm with small success probability ε.

2. Transform the classical algorithm into a quantum algorithm, for example with Grover search, with running time $O(p(n))$.

3. Apply the quantum amplitude amplification.

The result is a quantum algorithm with running time of $O(p(n)\frac{1}{\sqrt{\varepsilon}})$.

Now we consider two simple examples for the application of the quantum amplitude amplification:

Claw-finding: Given two integers M, N and two functions $f, g : [N] \rightarrow [M]$, find a pair $(x, y) \in [N]^2$, such that $f(x) = g(y)$.

A special case of claw-finding is the element distinctness problem, where $f, g : [N] \rightarrow \mathbb{N}$. There are two quantum algorithms for these problems. The first, due to Buhrman et al. [BDHHMSW01] uses Grover search and amplitude amplification, and solves this problems in $O(N^{3/4})$ quantum queries. The second, due to Ambainis [Amb04a], uses a quantum walk search (see Chapter 3.3) and finds a claw in $O(N^{2/3})$ quantum queries. This procedure is optimal, since there is a lower bound of $\Omega(N^{2/3})$ by Shi [AS04] resp. Kutin [Kut05].

Theorem 3.2.3 [BDHHMSW01] *The quantum query complexity of the claw-finding problem is $O(N^{3/4})$ and the quantum time complexity is $O(N^{3/4} \log^3 N)$.*

Proof. The claw-finding algorithm consists of the following three steps:

1. Choose a set $A \subset [N]$ of size r.

2. Use a classical algorithm for sorting the set $\{f(a) \mid a \in A\}$.

3. Find $y \in [N]$ such that $\exists x \in A$ with $f(x) = g(y)$ by Grover search.

Step 1 uses $O(r)$ queries. In step 2 we need no quantum queries, but $O(r \log r)$ time steps for sorting. Step 3 can be done in $O(\sqrt{N})$ quantum queries, and $O(\sqrt{N} \log r)$ quantum time steps, since testing if there is an $x \in A$ such that $f(x) = g(y)$ for a given $y \in [N]$ costs $O(\log r)$ comparisons. The success probability of this algorithm is $\varepsilon = r/N$, if there is a claw. Then the total quantum query complexity of claw-finding is $O((r + \sqrt{N}) \cdot \sqrt{N/r}) = O(N^{3/4})$ if $r = \sqrt{N}$. The quantum time complexity of this algorithm is $O(N^{3/4} \log^3 N)$, by using Remark 3.1.9. $\qquad\square$

Now we present a simple $O(\sqrt{nm})$ quantum query algorithm for finding a triangle in a graph $G = (V, E)$ with $n = |V|$ and $m = |E|$:

Triangle finding: Given a graph $G = (V, E)$, find distinct vertices $u, v, w \in V$ such that $\{u, v\}, \{v, w\}, \{u, w\} \in E$.

Theorem 3.2.4 [BDHHMSW01] *The quantum query complexity of the triangle finding problem is $O(n^{1.5})$ and the quantum time complexity is $O(n^{1.5} \log^2 n)$.*

Proof. The triangle finding algorithm consists of the following steps:

1. Find an edge $e = \{u, v\}$ in G with Grover search.

2. Find a vertex w with $\{u, v\}, \{v, w\}, \{u, w\} \in E$ with Grover search.

3. Apply amplitude amplification.

The step 1 takes $O(\sqrt{n^2/m})$ and step 2 takes $O(\sqrt{n})$ quantum queries to the adjacency matrix. If there is a triangle in the graph, then the probability that we find an edge belonging to this specific triangle is at least $3/m$. In total, triangle finding can be done in

$$O\left((\sqrt{n^2/m} + \sqrt{n})\sqrt{m}\right) = O(n + \sqrt{nm})),$$

resp. $O(n^{1.5})$ quantum queries if $m = O(n^2)$. $\qquad\square$

The best known quantum query lower bound for triangle finding is $\Omega(n)$ [BDHHMSW01]. There is an $O(n^{1.3})$ triangle finding algorithm by Magniez et al. [MSS05] which uses quantum walks (see Section 5.1). Therefore the quantum algorithm by Buhrman et al. [BDHHMSW01] is optimal for $m = O(n)$ and faster than the quantum walk algorithm for $m = O(n^{1.6})$.

3.3 Discrete Quantum Walk

Quantum walks are the quantum counterpart of Markov chains and random walks. A discrete quantum walk is a way of formulating local quantum dynamics on a graph. The walk takes discrete steps between neighbouring vertices, and it is a sequence of unitary transformations.

In this section we present the quantum walk search schema by Magniez et al. [MNRS07] for finding a marked element in the state space of a classical Markov chain. The quantum walk search provides a promising source for new quantum algorithms, like element distinctness algorithm [Amb04a], triangle finding [MSS05], testing group commutativity [MN05], matrix verification [BŠ06] and testing associativity [DT07].

Scheme for Searching

Let $P = (p_{xy})$ be the transition matrix of an ergodic symmetric Markov chain on the state space X with stationary distribution $\pi = (\pi_x)$. Let $M \subseteq X$ be a set of marked states. Assume that the search algorithms use a data structure D that associates some data $D(x)$ with every state $x \in X$. From $D(x)$, we would like to determine if $x \in M$. When operating on D, we consider the following three types of costs:

- *Setup cost* s: The worst case cost to compute $D(x)$, for $x \in X$.

- *Update cost* u: The worst case cost for transition from x to y, and update $D(x)$ to $D(y)$.

- *Checking cost* c: The worst case cost for checking if $x \in M$ by using $D(x)$.

More precise, suppose $0 \in X$, the setup cost is the cost for constructing the state

$$\sum_x \sqrt{\pi_x} |x, D(x)\rangle |0, D(0)\rangle,$$

The update cost is the cost to realize the unitary transformation and its inverse

$$|x, D(x)\rangle|0, D(0)\rangle \rightarrow |x, D(x)\rangle \sum_y \sqrt{p_{xy}}|y, D(y)\rangle$$

$$|0, D(0)\rangle|y, D(y)\rangle \rightarrow \sum_x \sqrt{p_{yx^*}}|x, D(x)\rangle|y, D(y)\rangle,$$

where $P^* = (p_{xy}^*)$ is the time-reversed Markov chain defined by the equations $\pi_x p_{xy} = \pi_y p_{yx}^*$. For checking we realize

$$|x, D(x)\rangle|y, D(y)\rangle \rightarrow -|x, D(x)\rangle|y, D(y)\rangle \text{ if } x \in M,$$

and leaves it unchanged otherwise.

Magniez et al. [MNRS07] developed a new scheme for quantum search, based on any ergodic Markov chain. Their work generalizes previous results by Ambainis [Amb04a] and Szegedy [Sze04a]. They extend the class of possible Markov chains and improve the query complexity as follows.

Theorem 3.3.1 [MNRS07] *Let $\delta > 0$ be the eigenvalue gap of an ergodic Markov chain P and let $\frac{|M|}{|X|} \geq \varepsilon$. Then there is a quantum algorithm that determines if M is empty or finds an element of M with cost*

$$s + \frac{1}{\sqrt{\varepsilon}}\left(\frac{1}{\sqrt{\delta}}u + c\right).$$

In the most practical applications (see [Amb04a, MSS05, BŠ06, DT07]) the quantum walk takes place on the Johnson graph which is defined as follows: the vertices are subsets of $\{1, \ldots, n\}$ of size r, and two vertices are connected iff they differ in exactly one number. It is well known that the spectral gap δ of $J(n,r)$ is $\Theta(1/r)$ for $1 \leq r \leq \frac{n}{2}$ (see e.g. [Knu03]).

Theorem 3.3.2 [Amb04a] *Let P be a random walk on the Johnson graph $J(n,r)$, where $r = o(n)$. Let M be either empty, or the class of all r subsets that contain a fixed subset of constant size $k \leq r$. Then, there is a quantum algorithm that determines if M is non-empty or finds a k-subset of M with cost*

$$\min_{k \leq r \leq n} \left\{ s + \left(\frac{n}{r}\right)^{k/2} \left(\sqrt{r}u + c\right) \right\}.$$

Proof. We apply Theorem 3.3.1. Suppose there is at least one marked k-subset, then there are $\binom{n-k}{r-k}$ marked vertices of the Johnson graph, and therefore

$$\varepsilon \geq \frac{\binom{n-k}{r-k}}{\binom{n}{r}} = \Omega\left(\frac{r^k}{n^k}\right).$$

$\square$

Corollary 3.3.3 *Let $P \subset [N]^k$ be a search problem with search space $[N]$, whereby P is associate by its characteristic function $f_P : [N]^k \to \{0,1\}$. The quantum query complexity for finding a subset $U \subset [N]$ of constant size k with $f_P(U) = 1$ is $O(N^{k/(k+1)} \cdot t)$, where t is the number of queries to the input for the computation of f_P.*

Proof. We apply the quantum walk search in the Johnson graph of Theorem 3.3.2. Let A be a subset of $[N]$ of size $r > k$. We will determine r later. A vertex A of the Johnson graph $J(N,r)$ is marked, if there is a subset $A' \subseteq A$ of size k with $f_P(A') = 1$. The database is the set $D(A) =$

$\{(i, d(i)) \mid i \in A\}$, where $d(i)$ is the part of the input corresponding to the i'th element in the search space which we need to verify if the vertex A is marked. The setup cost for the database $D(A)$ is rt and the update cost is t. To check whether there are $i_1, \ldots, i_k \in A$ such that $(i_1, \ldots, i_k) \in P$ requires no queries. Then the quantum query complexity for finding a k-subset is

$$O\left(r \cdot t + \left(\frac{N}{r}\right)^{k/2} \cdot \sqrt{r} \cdot t\right).$$

This expression is minimal for $r = N^{k/(k+1)}$, therefore we have $O(N^{k/(k+1)}t)$. $\qquad\square$

Now we present several applications of the quantum walk search.

Grover search. Grover [Gro96] discovered a quantum search algorithm, which is quadratic faster than a classical search. By using Corollary 3.3.3, we can implement this algorithm by a quantum walk in the Johnson graph. Then the quantum query complexity for finding a solution x of a search problem $P \subset [N]$ is $O(\sqrt{N} \cdot t)$, where t is the number of queries for the verification if x is a solution of P.

Element distinctness. Ambainis [Amb04a] presented the first algorithm which uses quantum walk search and that went beyond the capability of Grover search. He considered the element distinctness problem: Given N numbers $x_1, \ldots, x_N$, find two distinct indices $i, j \in [N]$ with $x_i = x_j$. By using Theorem 3.3.3, we search for a tupel $(i, j) \in [N]^2$ with $(i, j) \in P$ iff $x_i = x_j$. Then the element distinctness problem can be solved in $O(N^{2/3})$ quantum queries, which is optimal (see [AS04, Kut05]). The quantum query complexity for finding k equal numbers is $O(N^{k/(k+1)})$. The best lower bound for this general case is only $\Omega(N^{2/3})$ (see [Amb04a]).

Matrix Verification. Buhrman and Špalek [BŠ06] constructed an $O(n^{5/3})$ quantum algorithm for the matrix verification problem. We have given three $n \times n$ matrices A, B, C, one has to decide whether $AB = C$. Their algorithm uses the quantum walk by Szegedy [Sze04a] in the graph categorical product of two Johnson graphs. By using Corollary 3.3.3 we can simplify their algorithm. We search for a tupel $(i, j) \in [n]^2$ with $(i, j) \in P$ iff $A_{i,*} \cdot B_{*,j} - C_{i,j} \neq 0$. Then $k = 2$ and $t = O(n)$, since we need $O(n)$ queries to the input matrices for the verification if $(i, j) \in P$.

Triangle finding. Magniez, Santha and Szegedy [MSS05] presented an $O(n^{1.3})$ quantum algorithm for triangle finding. They used the quantum walk search of Theorem 3.3.2 for finding two vertices $u, v \in V$, such that $\{u, v\}$ is an edge of a triangle in G. The database is the vertex-induced subgraph $G[A]$, where $A \subset V$ of size r. The setup cost is r^2 to determine $G[A]$ and the update cost is r. For checking if a vertex A in the Johnson graph $J(n, r)$ is marked, we have to search for a vertex $w \in V$, such that there is an edge $\{u, v\}$ in $G[A]$ which forms a triangle with w. This step can be done with Grover search and a special form of element distinctness procedure by Ambainis [Amb04a], therefore the checking cost is $O(\sqrt{n}r^{2/3})$. Then the quantum query complexity of triangle finding is

$$r^2 + \frac{n}{r}\left(\sqrt{r}r + \sqrt{n}r^{2/3}\right) = O(n^{13/10}) \text{ for } r = n^{3/5}.$$

Graph copy. The triangle finding algorithm can be generalised for finding in a graph G a copy of a given graph H. Magniez, Santha and Szegedy [MSS05] showed that the quantum query complexity for finding such a copy is $O(n^{2-2/k})$, where k is the number of vertices H with $k > 3$.

Quantization of Markov Chains

In this subsection we define the quantum analogue of an irreducible Markov chain. Let $P = (p_{xy})$ be the transition matrix of any irreducible Markov chain on a finite space X of size n. By the Perron-Frobenius theorem, the chain has a unique stationary distribution $\pi = (\pi_x)$. Let $P^* = (p_{yx}^*)$ be the time-reversed Markov chain defined by the equations $\pi_x p_{xy} = \pi_y p_{yx}^*$. The Markov chain P is said to be *reversible* if $P^* = P$.

Let I denote the identity operator on Hilbert space $\mathcal{H}$. For a state $|\psi\rangle \in \mathcal{H}$, let $\Pi_\psi = |\psi\rangle\langle\psi|$ be the orthogonal projector on $\mathrm{Span}(|\psi\rangle)$, and $\mathrm{ref}(\psi) = 2\Pi_\psi - I$ the reflection through the line generated by $|\psi\rangle$. Let $\mathcal{K}$ be a subspace of $\mathcal{H}$, and $\mathrm{ref}(\mathcal{K})$ the respective reflections through subspaces $\mathcal{K}$.

We denote with $\mathcal{A} = \mathrm{Span}(|x\rangle|p_x\rangle \mid x \in X)$ and $\mathcal{B} =$

$\text{Span}(|p_y^*\rangle|y\rangle \mid y \in X)$ vector subspaces of $\mathcal{H} = \mathbb{C}^{X \times X}$, where

$$|p_x\rangle = \sum_{y \in X} \sqrt{p_{xy}} \, |y\rangle \quad \text{and} \quad |p_y^*\rangle = \sum_{x \in X} \sqrt{p_{yx}^*} \, |x\rangle.$$

Definition 3.3.4 The unitary transformation

$$W(P) := \text{ref}(\mathcal{B}) \cdot \text{ref}(\mathcal{A})$$

is called the *quantum walk* based on the classical Markov chain P.

Definition 3.3.5 Let P be a Markov chain, the *discriminant* matrix is defined by

$$D(P) := \left(\sqrt{\tfrac{\pi_x}{\pi_y}} \, p_{xy} \right) = \text{diag}(\pi)^{1/2} \cdot P \cdot \text{diag}(\pi)^{-1/2},$$

where $\text{diag}(\pi)$ is the invertible diagonal matrix with the coordinates of the distribution π in its diagonal.

The singular values of $D(P)$ (roots of eigenvalues of $D(P)^T D(P)$) all lie in the range $[0, 1]$, we may express them as $\cos\theta$, for some angles $\theta \in [0, \tfrac{\pi}{2}]$.

Let $\Delta(P)$ be the *phase gap* of $W(P)$ which is 2θ, where θ is the smallest angle in $(0, \tfrac{\pi}{2})$ such that $\cos\theta$ is a singular value of $D(P)$. The angular distance of 1 from any other eigenvalue is at least $\Delta(P)$.

Proposition 3.3.6 [MNRS07] *Let P be an ergodic reversible classical Markov chain, then:*

1. *The singular values and absolute values of its eigenvalues of $D(P)$ are equal.*

62

2. *The spectra of P and $D(P)$ are the same, and the eigenvalue 1 is the only eigenvalue of P with absolute value 1.*

3. *The eigenvector of $D(P)$ with eigenvalue 1 is $\pi = (\sqrt{\pi_x})$, and every singular (or eigen-) vector of $D(P)$ orthogonal to this has singular value strictly less than 1.*

4. *The operator $W(P)$ has a unique eigenvector $|\pi\rangle$ in $\mathcal{A} + \mathcal{B}$ with eigenvalue 1, and the remaining eigenvalues in $\mathcal{A} + \mathcal{B}$ are bounded away from 1.*

Quantum Walk Search Technique

In this subsection we present the quantum walk search algorithm by Magniez et al. [MNRS07]. The main idea of this search is the application of the quantum phase estimation [CEMM98] to the quantum walk in order to implement an approximate reflection operator. This operator is then used in an amplitude amplification [BHMT02] scheme.

The work by [MNRS07] generalizes previous results by Ambainis [Amb04a] and Szegedy [Sze04a]. They extend the class of possible Markov chains and improve the quantum complexity. The general ideas of the quantum walk search are the following:

1. Let $\mathcal{H} := \mathbb{C}^{X \times X}$ be the Hilbert space of the algorithm, and

$$|\pi\rangle = \sum_{x \in X} \sqrt{\pi_x} |x\rangle |p_x\rangle$$

the initial state, which corresponds to starting in the stationary distribution π.

2. Assume that there is at least one marked element, let $\mathcal{M} = \mathbb{C}^{M \times X}$ denote the subspace with marked items in the first register, and let

$$\Pi_{\mathcal{M}} := \sum_{x \in M, y \in X} |xy\rangle\langle xy|$$

be the projector onto this subspace.

3. The task is to transform the initial state $|\pi\rangle$ to the target state $|\mu\rangle$, which is the (normalized) projection of $|\pi\rangle$ onto the "marked subspace" $\mathcal{M}$. The state $|\mu\rangle$ is given by

$$|\mu\rangle = \frac{\Pi_{\mathcal{M}}|\pi\rangle}{||\Pi_{\mathcal{M}}|\pi\rangle||} = \frac{1}{\sqrt{\varepsilon}} \sum_{x \in M} \sqrt{\pi_x} |x\rangle|p_x\rangle,$$

where $\varepsilon := ||\Pi_{\mathcal{M}}|\pi\rangle||^2 = \sum_{x \in M} \pi(x)$ is the probability of a set M of marked states under the stationary distribution π.

4. The transformation is implemented by a Grover rotation in the two-dimensional subspace spanned by the states $|\pi\rangle$ and $|\mu\rangle$. The Grover operator is the transformation

$$\mathrm{ref}(\pi) \cdot \mathrm{ref}(\mu^{\perp}),$$

where $|\mu^{\perp}\rangle$ is orthogonal to $|\mu\rangle$. From the Grover algorithm [Gro96] follows that after $O(1/\sqrt{\varepsilon})$ iterations of this rotation starting with the state $|\pi\rangle$, we will have approximated the target state $|\mu\rangle$.

5. For the implementation of the $\mathrm{ref}(\pi)$ operator we use the phase estimation procedure on $W(P)$.

6. For the implementation of the ref($\mu^{\perp}$) operator, we check whether an item in the first register is marked, which incurs some checking cost c.

The main idea in the implementation of ref(π) is to use phase estimation with the quantum walk operator $W(P)$.

Theorem 3.3.7 [MNRS07] *Let P be an ergodic Markov chain on a state space of size n, and let $m = n^2$. Then there is a constant $c > 0$ such that for any integer k, there exists a quantum circuit $R(P)$ that acts on $m+ks$ qubits, where $s = \log(\frac{1}{\Delta(P)}) + O(1)$, and satisfies the following properties:*

1. *The circuit $R(P)$ uses $2ks$ Hadamard gates, $O(ks^2)$ controlled phase rotations, and makes at most $k\,2^{s+1}$ calls to the controlled quantum walk $W(P)$, and its inverse.*

2. *If $|\pi\rangle$ is the unique 1-eigenvector of $W(P)$ as defined above, then $R(P)|\pi\rangle|0^{ks}\rangle = |\pi\rangle|0^{ks}\rangle$.*

3. *If $|\psi\rangle$ lies in the subspace of $\mathcal{A} + \mathcal{B}$ orthogonal to $|\pi\rangle$, then*

$$||(R(P) + I)|\psi\rangle|0^{ks}\rangle|| \le 2^{1-ck}, \;\; i.e. \;\; R(P)|\psi\rangle \approx -|\psi\rangle.$$

Moreover the family of circuits $R(P)$ parametrized by n and k is uniform.

Now we describe the quantum walk search algorithm.

Algorithm 4 QUANTUM WALK SEARCH

Input: Markov chain P with state space X.

Output: Marked state $x \in M$ (if there is one).

1: Initialisation:
$$|\pi\rangle = \sum_x \sqrt{\pi(x)}|x, D(x)\rangle|p_x\rangle$$

2: Repeat $O(1/\sqrt{\varepsilon})$ times:

    1. Phase flip:

$$|x, D(x)\rangle|y, D(y)\rangle \rightarrow (-1)^{\chi_M(x)}|x, D(x)\rangle|y, D(y)\rangle$$

    2. Apply circuit $R(P)$ with $k = O(\log(1/\sqrt{\varepsilon}))$

3: Measure the first register

4: Output x, if $x \in M$; otherwise output 'no marked element exists'.

Theorem 3.3.8 [MNRS07] *Let $\delta > 0$ be the eigenvalue gap of a reversible, ergodic Markov chain P, and let $\varepsilon > \frac{|M|}{|X|}$. Then the* QUANTUM WALK SEARCH *algorithm determines if M is empty or finds an element of M with cost*

$$s + \frac{1}{\sqrt{\varepsilon}}\left(\frac{1}{\sqrt{\delta}}\log\frac{1}{\sqrt{\varepsilon}}u + c\right).$$

Proof. The algorithm rotates the state $|\pi\rangle$ near the target state $|\mu\rangle$ by using amplitude amplification with $O(1/\sqrt{\varepsilon})$ invocations. We prove that the QUANTUM WALK SEARCH algorithm simulates with an arbitrarily small error the GROVER ALGORITHM, and therefore finds a marked element with high probability, if such an element exists.

For $i \geq 0$, we define $|\phi_i\rangle$ as the result of i Grover iterations applied to $|\pi\rangle$, and $|\psi_i\rangle$ as the result of i iterations of step 2 in QUANTUM WALK

SEARCH applied to $|\pi\rangle$. It is not difficult to show by induction on i that $|||\psi_i\rangle - |\phi_i\rangle|| \leq i 2^{1-ck}$. This implies that $|||\psi_k\rangle - |\phi_k\rangle||$ is an arbitrarily small constant when $k = O(c^{-1}\log(1/\sqrt{\varepsilon}))$.

Let $\lambda_0, ..., \lambda_{n-1}$ be the eigenvalues of P such that $1 = \lambda_0 > |\lambda_1| \geq ... \geq |\lambda_{n-1}|$ and the spectral gap $\delta(P) = 1 - |\lambda_1| > 0$. Since the discriminant $D(P)$ is similar to P, their spectra are the same, and therefore the singular values of $D(P)$ are $|\lambda_0|, |\lambda_1|, ...$ By definition, $\Delta(P) = 2\theta_1$, where $\cos\theta_1 = |\lambda_1|$, then

$$\Delta(P) \geq |1 - e^{2i\theta_1}| = 2\sqrt{1 - |\lambda_1|^2} \geq 2\sqrt{\delta(P)}.$$

The total number of cost in each iteration is

$$\sqrt{\frac{1}{\varepsilon}} \cdot k \cdot 2^s = \frac{1}{\sqrt{\varepsilon\delta}} \cdot \log\frac{1}{\sqrt{\varepsilon}}.$$

$\square$

Remark 3.3.9 Magniez et al. [MNRS07] showed how an approximate reflection operator derived from a quantum walk may be incorporated into a search algorithm without incurring additional cost for reducing its error.

Chapter 4

Methods for Quantum Lower Bounds

In this Chapter we present several tools for the computation of quantum query lower bounds. First we present the quantum adversary method, which was initiated by Ambainis [Amb02]. This tool is very useful to prove lower bounds for the bounded error quantum query complexity of graph and algebra problems. The second method is the polynomial method, which was introduced by Beals et al. [BBCMW01]. The quantum query lower bound follows from a lower bound on the degree of a polynomial. We present some facts about the polynomial method in connection with some general lower bounds for graph problems.

The polynomial method and the adversary method are generally incomparable. There are examples which show that the polynomial method gives better bounds than the quantum adversary and vice versa.

4.1 Adversary Method

An important method to prove lower bounds for the bounded error quantum query complexity of Boolean functions is the quantum adversary method. The original version of the quantum adversary method is called unweighted method and was invented by Ambainis [Amb02]. This method starts with choosing a set of pairs of inputs, on which the Boolean function f takes different values. Then the quantum query lower bound for computing f is determined by some combinatorial properties of the input.

First we present a special case of the unweighted method, which is in many cases sufficient to prove tight lower bounds for the quantum query complexity of graph and algebra problems.

Theorem 4.1.1 [Amb02] *Let $A \subseteq \{0,1\}^n$, $B \subseteq \{0,1\}^n$ and $f : \{0,1\}^n \rightarrow \{0,1\}$ such that $f(x) = 1$ for all $x \in A$, and $f(y) = 0$ for all $y \in B$. Let m and m' be numbers such that*

1. *for every $(x_1, \ldots, x_n) \in A$ there are at least m values $i \in \{1, \ldots, n\}$ such that $(x_1, \ldots, x_{i-1}, 1 - x_i, x_{i+1}, \ldots, x_n) \in B$,*

2. *for every $(x_1, \ldots, x_n) \in B$ there are at least m' values $i \in \{1, \ldots, n\}$ such that $(x_1, \ldots, x_{i-1}, 1 - x_i, x_{i+1}, \ldots, x_n) \in A$.*

Then every bounded-error quantum algorithm that computes f has quantum query complexity $\Omega(\sqrt{m \cdot m'})$.

Example 4.1.2 We can use Theorem 4.1.1 to prove the $\Omega(\sqrt{N})$ quantum query lower bound for quantum search. Suppose we have an array of Boolean values of size N. We show that deciding whether it contains an entry with value one requires $\Omega(\sqrt{N})$ quantum queries. The set A consists of all arrays of length N, which contains exactly one entry with value one. The set B consists of the zero array of length N. From each $x \in A$, we can obtain $x' \in B$ by changing the entry from 1 to 0, then $m = 1$. From each $x' \in B$, we can obtain $x \in A$ by changing one entry from 0 to 1, then $m' = n$. Therefore the query complexity of quantum search is $\Omega(\sqrt{N})$.

For some special problems, we need a stronger version than Theorem 4.1.1 for proving quantum query lower bounds.

Theorem 4.1.3 [Amb02] *Let* $A \subseteq \{0,1\}^n, B \subseteq \{0,1\}^n$ *and* $f : \{0,1\}^n \to \{0,1\}$ *such that* $f(x) = 1$ *and* $f(y) = 0$ *for all* $x \in A$ *and* $y \in B$. *Let* $R \subset A \times B$ *and* m, m', l, l' *be numbers such that*

1. *for every* $x \in A$, *there are at least* m *different* $y \in B$ *such that* $(x, y) \in R$.

2. *for every* $y \in B$, *there are at least* m' *different* $x \in A$ *such that* $(x, y) \in R$.

3. *for every* $x \in A$ *and* $i \in [n]$, *there are at most* l *different* $y \in B$ *such that* $(x, y) \in R$ *and* $x_i \neq y_i$.

4. *for every* $y \in B$ *and* $i \in [n]$, *there are at most* l' *different* $x \in A$

such that $(x, y) \in R$ and $x_i \neq y_i$.

Then every bounded-error quantum algorithm that computes f has quantum query complexity $\Omega\left(\sqrt{\frac{m \cdot m'}{l l'}}\right)$.

Remark 4.1.4

1. If we choose $(x, y) \in R$ iff x and y differ in exactly one position, then $l = l' = 1$, and we get Theorem 4.1.1 as a special case of Theorem 4.1.3.

2. We can also prove quantum query lower bounds for general functions or matrices. Let $\mathcal{F} = \{f : [n] \times [n] \to [n]\}$ be the set of all possible input function and $\Phi : \mathcal{F} \to \{0, 1\}$. Then $A, B \subset \mathcal{F}$ such that $\Phi(f) = 1$ and $\Phi(g) = 0$ for all $f \in A$ and $g \in B$. The lower bound for the computation of Φ follows similar to Theorem 4.1.3. For further application, we define

$$d(f, g) := |\{\, x, y \in [n] \mid f(x, y) \neq g(x, y) \,\}|.$$

4.2 Polynomial Method

The polynomial method for proving quantum query lower bounds was introduced by Beals et al. [BBCMW01]. This method is based on the observation that the measurement probabilities can be described by low degree polynomials in the input bits. It was shown that, if t queries have been done, then the degree of the polynomials is at most $2t$.

The polynomial method has been successfully applied to obtain tight lower bounds for the collision problem and the element distinctness problem (see [AS04][Kut05]). The quantum adversary method is unable to prove such lower bounds. With the polynomial method we can also prove lower bounds for the exact and zero-error quantum complexity. The adversary method completely fails in this setting, here we get only bounded-error lower bounds. On the other hand, the adversary method gives for some functions better lower bounds than the polynomial method. The biggest proved gap between the two methods is $n^{1.321}$ (see [Amb03]). The polynomial method did not yet succeed in proving lower bounds that are very simple to prove by the adversary method.

Definition 4.2.1 Let $f : \{0,1\}^N \rightarrow \{0,1\}$ be a Boolean function and $p : \mathbb{R}^N \rightarrow \mathbb{R}$ be an N-variate polynomial. If $p(x) = f(x)$ for all $x \in \{0,1\}^N$, then we say p *represents* f. We denote by $deg(f)$ the degree of a minimum-degree polynomial p that represents f. If $|p(x) - f(x)| \leq 1/3$ for all $x \in \{0,1\}^N$, we say p *approximates* f, and $\widetilde{deg}(f)$ denotes the degree of a minimum-degree p that approximates f.

By using the quantum query model, it is not difficult to show the following fact:

Lemma 4.2.2 [BBCMW01] Let $|\varphi\rangle = \sum_x \alpha(x_1, \ldots, x_N) |x\rangle$ be the state of a quantum algorithm after T queries. Then $\alpha(x_1, \ldots, x_N)$ being polynomials of degree at most T, and the probability of the measurement of value x is $\alpha(x_1, \ldots, x_N)^2$.

Then we can write the acceptance probability of a quantum algorithms with T queries as a polynomial $p(x)$ of degree $\leq 2T$, therefore we have

Theorem 4.2.3 [BBCMW01]

$$Q_E(f) \geq deg(f)/2 \quad \text{and} \quad Q_2(f) \geq \widetilde{deg}(f)/2.$$

A simple tool for proving lower bounds can be done by computing the sensitivity of a Boolean function.

Definition 4.2.4 The *sensitivity* of $f : \{0,1\}^N \to \{0,1\}$ on input x is

$$s_x(f) = |\{i \mid f(x) \neq f(x \oplus e_i)\}|,$$

where e_i is the i'th unit vector. The *average sensitivity* of f is

$$\overline{s}(f) = \frac{1}{2^N} \sum_{x \in \{0,1\}^N} s_x(f).$$

Beals et al. [BBCMW01] showed that

$$\Omega(\overline{s}(f)) = \widetilde{deg}(f)$$

Example 4.2.5 In combination with Theorem 4.2.3 it is not difficult to see for example that the computation of the parity function $f_p(x_1, \ldots, x_N) = x_1 \oplus \ldots \oplus x_N$ requires $\Omega(N)$ quantum queries, since $s_x(f_p) = n$ for all $x \in \{0,1\}^N$ and therefore $\overline{s}(f_p) = n$.

Now we are interested in the number of queries to the adjacency matrix, which we need to determine if a given graph has a certain property. By using the polynomial method we can prove lower bounds for the exact, zero error and bounded error quantum query complexity of such general graph problems.

Definition 4.2.6 A *graph property* P is a subset of the set of all graphs that is closed under permutation of the nodes, i.e. if G_1, G_2 represent isomorphic graphs, then $G_1 \in P$ iff $G_2 \in P$. A graph property called *monotone*, if adding edges cannot destroy the property.

Example 4.2.7 Let $G = (V, E)$ be a graph, a monotone graph property is for example: "G contains a clique of size k", since if we add edges in G we cannot destroy the clique. On the other hand, the property "G contains a independent set of size k" is not monotone.

Theorem 4.2.8 [BCWZ99] *For all monotone graph properties P hold:*

$$Q_E(P) \in \Omega(n^2), \; Q_0(P) \in \Omega(n) \text{ and } Q_2(P) \in \Omega(\sqrt{n}).$$

Furthermore, there exists graph properties P_1, P_2, P_3, such that $Q_E(P_1) < n(n-1)$ for every $n > 2$, $Q_0(P_2) \in O(n^{3/2})$ and $Q_2(P_3) \in O(n)$.

The complexity of monotone graph properties has been well-studied classically. Researchers suppose that for every monotone graph property P, it holds that $D(P) = n(n-1)$, see [LY94]. This is called the Aanderaa-Karp-Rosenberg conjecture and is still open. The best known general bound is $D(P) \in \Omega(n^2)$, see [Kin88]. For the classical zero-error complexity, the best known general result is $R_0(P) \in \Omega(n^{4/3})$ [Haj91], but it has been conjectured that $R_0(P) \in \Theta(n^2)$.

For the bounded error quantum query complexity, we can easily verify that $Q_2(P) \in \Omega(\sqrt{n})$, since we have quadratic gaps between quantum and classical complexity. For the property $P =$ "the graph has at least one edge" holds $Q_2(P) \in O(n)$ by quantum search. Combining this with $D(P) \in \Omega(n^2)$ and $D(f) \in O(Q_2(f)^4)$ for all monotone Boolean functions f (see [BBCMW01]), we obtain the general lower bound.

Chapter 5

Fundamental Graph Problems

In this Chapter we present fundamental quantum graph algorithms. First we give quantum algorithms for basic graph problems, like depth first search, breadth first search and topological numbering. Then we present a uniform and simplified description of known quantum graph algorithms. We give the quantum algorithms by Dürr et al. [DHHM04] for the minimum spanning tree, graph connectivity, strong graph connectivity and the single source shortest path problem. Furthermore we present a quantum algorithm by Ambainis and Špalek [AS06] for determining a maximum flow in a directed graph with bounded edge weight. We use some of these quantum algorithms as a black box to speed up other graph algorithms.

5.1   Basic Graph Algorithms

For many quantum graph algorithms in this thesis we use Grover's algorithm [Gro96] for searching of edges in the graph. The following proposition gives the quantum query complexity of this search:

Proposition 5.1.1 *Given an undirected graph* $G = (V, E)$.

1. *The existence of an edge* $\{u, v\}$ *in* G *can be tested by a single query to the adjacency matrix and* $O(\sqrt{\min\{d_G(u), d_G(v)\}})$ *quantum queries to the adjacency list of* G.

2. *All neighbours of a vertex* v *of* G *can be found in* $O(\sqrt{n \cdot d_G(v)})$ *quantum queries to the adjacency matrix of* G.

3. *All neighbours of a vertex* v *of* G *with a special property can be found in* $O(\sqrt{d_G(v) \cdot a_v})$ *quantum queries to the adjacency list, where* a_v *is the number of adjacent vertices of* v *with the special property.*

Proof. The three facts follows immediately from Theorem 3.1.8. $\qquad \square$

Many graph algorithms use depth first search, breadth first search and topological numbering as subroutines. Classically these searches can be done in linear time in the number of edges of the graph. With the application of quantum search, we can speed up these subroutines.

Depth First Search (DFS): Given a graph $G = (V, E)$ and a vertex $s \in V$, compute a depth-first tree T from s, such that T is rooted at s and contains all the vertices of G that are reachable from s.

Lemma 5.1.2 *The quantum query complexity of DFS is* $O(n^{1.5} \log n)$ *in the adjacency matrix model and* $O(\sqrt{nm} \log n)$ *in the adjacency list model.*

Proof. In DFS, we construct a tree T that contains all the vertices of G that are reachable from the root s. We can construct this tree by the following simple procedure: For each edge $(s, w) \in E$ in which w has not been discovered by T, make w the next child of s in T, and recall the procedure with $s = w$.

We use the Grover search to construct the tree T. Every vertex is discovered by T at most once. In the adjacency matrix model, every vertex is found in $O(\sqrt{n})$ quantum queries. In total we are using $O(n^{1.5})$ quantum queries in the adjacency matrix model. In the list model, finding an adjacent vertex of s which has not been discovered uses $O(\sqrt{d_G(s)})$ quantum queries. In total we have $\sum_s \sqrt{d_G(s)} = O(\sqrt{mn})$ quantum queries in the adjacency list model. In order to get a constant success probability, we need to amplify the success probability of each subroutine by repeating it $O(\log n)$ times, see Remark 3.1.9. $\qquad\square$

Breadth First Search (BFS): Given a graph $G = (V, E)$ and a vertex $s \in V$, compute a breadth-first tree T from s, such that T is rooted at s and contains all the vertices of G that are reachable from s.

Lemma 5.1.3 *The quantum query complexity of BFS is $O(n^{1.5} \log n)$ in the adjacency matrix model and $O(\sqrt{nm} \log n)$ in the adjacency list model.*

Proof. In BFS, we construct a tree T that contains all the vertices of G that are reachable from the root s by the following simple procedure: Let $V_0 = \{s\}$ and $i = 0$. While $V_i \neq \emptyset$ we construct the set V_{i+1}: First we set $V_{i+1} = \emptyset$, and for each $v \in V_i$ and each edge (v, w), make w the next child of v in T, if w has not been discovered by T. We add w to V_{i+1} and repeat this loop with $i = i + 1$.

In the adjacency matrix model a vertex is found in $O(\sqrt{n})$ quantum queries. Every vertex is processed at most once. In the list model, processing a vertex v costs $O(\sqrt{d_G(v) \cdot n_v})$ quantum queries, where n_v is the number of adjacent vertices to v, which has not been discovered. Since $\sum_v n_v \leq n$, then the total quantum query complexity is upper-bounded by the Cauchy-Schwarz inequality:

$$\sum_v \sqrt{d_G(v) \cdot n_v} \leq \sqrt{\sum_v d_G(v)} \sqrt{\sum_v n_v} = O(\sqrt{mn}).$$

By using Remark 3.1.9, the error probability of this algorithm is constant.

$\square$

Topological Numbering (TN): Given a directed acyclic graph $G = (V, E)$, compute a numbering $I : V \to [n]$, such that each edge is directed from lower number to higher number, i.e. if $(u, v) \in E$ then $I(u) < I(v)$.

Lemma 5.1.4 *The quantum query complexity of topological numbering is $O(n^{1.5} \log n)$ in the adjacency matrix model and $O(\sqrt{nm} \log n)$ in the adjacency list model.*

Proof. We grow a DFS path P until a sink t (vertex with outdegree 0) is reached. Then we set $I(t) = n$, decrease n by 1 and delete t from the path P and the graph G. We continue with the DFS procedure until G

has no vertices.

In each iteration we grow the DFS path P by starting with the previous P and extending it, if possible. Since we use DFS, we apply Lemma 5.1.2 to obtain the quantum query complexity of topological numbering.

$\square$

Corollary 5.1.5 *The quantum time complexity of BFS, DFS and TN is* $O(n^{1.5} \log^2 n)$ *in the adjacency matrix model and* $O(\sqrt{nm} \log^2 n)$ *in the adjacency list model.*

5.2 Graph Connectivity

In this Section we present a short description of the quantum query algorithms by Dürr et al. [DHHM04] which decide if an undirected resp. directed graph is connected. At first we consider the connectivity problem for undirected graphs.

Graph Connectivity: Given an undirected graph G, decide if $G = (V, E)$ is connected.

We compute a spanning tree, provided the graph G is connected. First we consider the adjacency matrix model. The algorithms starts with n connected components, one for each vertex. Then we construct a spanning tree by repeatedly choosing an edge that connects two of the components. Let T be a subgraph of G, for the application of quantum search, we define a search function $f_T : E \rightarrow \{0, 1\}$ with

$$f_T(e) := \begin{cases} 1, & \text{if } c(T_{+e}) < c(T) \\ 0, & \text{otherwise,} \end{cases}$$

where $c(G)$ is the number of connected components of the graph G. The quantum query algorithm for the graph connectivity problem in the adjacency matrix model is the following:

Algorithm 5 GRAPH CONNECTIVITY-M

Input: Undirected graph $G = (V, E)$ in the adjacency matrix model.

Output: Spanning tree $T = (V, A)$, if G is connected.

Complexity: M: $O(n^{1.5})$ quantum queries.

1: $A := \emptyset$, $T := (V, A)$
2: **while** $c(T) > 1$ **do**
3:    $e := $ QUANTUM SEARCH$[f_T]$
4:    $A := A \cup \{e\}$

Theorem 5.2.1 [DHHM04] *The expected quantum query complexity of the* GRAPH CONNECTIVITY-M *algorithm is* $O(n^{1.5})$ *in the adjacency matrix model.*

Proof.   The algorithm searches an edge that connects two different components in T. This step can be done in $O(\sqrt{n^2/k})$ quantum queries, if there are $k > 0$ edges that connect two different components in A. Suppose the graph is connected, then there are exactly $n - 1$ iterations, and the expected total number of quantum queries is

$$\sum_{k=2}^{n} \sqrt{\frac{n^2}{k-1}} = O(n^{1.5}).$$

We may choose to stop the algorithm after twice the expected number of quantum queries, then we have bounded one-sided error algorithm.

□

Theorem 5.2.2 [DHHM04] *The graph connectivity problem requires* $\Omega(n^{1.5})$ *quantum queries to the adjacency matrix.*

Proof. We use Theorem 4.1.3. Let A be the set of all graphs $G = (V, E)$ with n vertices and an unique cycle. Let B be the set of all graphs $G' = (V, E')$ with n vertices and exactly two cycles, each of length between $n/3$ and $2n/3$. We define the relation $(G, G') \in R$, if there exist four vertices $v_1, v_2, v_3, v_4 \in V$ such that the only difference between G and G' is that $(v_1, v_2), (v_3, v_4) \in E$ but not in E' and $(v_1, v_3), (v_2, v_4) \in E'$ but not in E. Then it holds $m = O(n^2)$, since there are n choices for the first edge and $n/3$ choices for the second edge. Analog $m' = O(n^2)$, since from each cycle one edge must be chosen. We have $l_{G,(i,j)} = 4$, iff $(i, j) \notin E$, since in G' we have the additional edge (i, j) and the endpoints of the second edge must be neighbours of i and j. On the other hand $l_{G,(i,j)} = O(n)$ iff $(i, j) \in E$ since then (i, j) is one of the edges to be removed and there remains $n/3$ choices for the second edge. The values $l_{G',(i,j)}$ are similar, such that $l_{\max} = O(n)$. Then the quantum query complexity of graph connectivity is $\Omega(\sqrt{l \cdot l'/l_{\max}}) = \Omega(n^{1.5})$. □

Now we determine the quantum query complexity of the graph connectivity problem in the adjacency list model. We need the following simple fact from graph theory:

Lemma 5.2.3 [DHHM04] *Let* $G = (V, E)$ *be a graph in the adjacency list model. The construction of a partition of* V *into a set of connected components* $\{C_1, \ldots, C_k\}$ *for some integer* k *with the property*

$$m_{C_j} := \sum_{i \in C_j} d_G(i) \leq |C_j|^2, \ \forall j \in [k],$$

can be done in $O(n)$ *classical queries.*

We denote with CONNECTED COMPONENTS$[G]$ the classical algorithm of Lemma 5.2.3, which computes a set of connected components $\{C_1, \ldots, C_k\}$. Let T be a subgraph of $G = (V, E)$, $C \subseteq V$ and $f_{G,T,C} : E \to \{0, 1\}$ a search function with

$$f_{G,T,C}(e) := \begin{cases} 1, & \text{if } c(T_{+e}) < c(T) \text{ and } e \in C \times N_G(C) \\ 0, & \text{otherwise.} \end{cases}$$

Algorithm 6 GRAPH CONNECTIVITY-L

Input: Undirected graph $G = (V, E)$ in the adjacency list model.

Output: Spanning tree T, if G is connected.

Complexity: L: $O(n)$ quantum queries.

  1: $\mathcal{C} := $ CONNECTED COMPONENTS$[G]$

  2: $A := \emptyset$, $T := (V, A)$

  3: **while** $c(T) > 1$ **do**

  4:     Choose $C \in \mathcal{C}$ with $m_C = \min\{m_{C_1}, \ldots, m_{C_k}\}$

  5:     $e := $ QUANTUM SEARCH$[f_{G,T,C}]$

  6:     $A := A \cup \{e\}$

Theorem 5.2.4 [DHHM04] *The expected quantum query complexity of the* GRAPH CONNECTIVITY-L *algorithm is* $O(n)$ *in the adjacency list model.*

Proof. Suppose that the graph G is connected. First we construct a set of k connected components of Lemma 5.2.3 in $O(n)$ classical queries. In every iteration of the algorithm, we choose a component with smallest total degree. Then we use Grover's algorithm for searching an edge out of C. This step can be done in $O(\sqrt{m_C}) \leq O(|C|)$ quantum queries to the adjacency list. Summing over all k components, the total number of expected quantum queries is $O(n)$. $\qquad\square$

The quantum query lower bound of $\Omega(n)$ for the graph connectivity in the adjacency list model can be proved by a straightforward reduction from parity, see [DHHM04].

Now we consider the graph connectivity problem for directed graphs. We simplify the analysis of the strong connectivity algorithm by Dürr et al. [DHHM04].

Strong Graph Connectivity: Given a directed graph G, decide if there is a directed path between every pair of vertices of G.

Theorem 5.2.5 [DHHM04] *The quantum query complexity of the strong graph connectivity problem is $O(n^{1.5} \log n)$ in the adjacency matrix model and $O(\sqrt{nm} \log n)$ in the adjacency list model.*

Proof. The algorithm consists of two steps: First we construct a directed depth first spanning tree $T = (V, A)$, where the vertices are labeled according to the order which they are added to T. Second we search for every vertex of V the neighbour with smallest label. The result of this search is a set of edges $B \subseteq E$. The following fact can be showed:

Claim: $G = (V, E)$ is strongly connected iff $G' = (V, A \cup B)$ is strongly connected.

In the first step, the set A can be computed in $O(\sqrt{n^{1.5}} \log n)$ resp. $O(\sqrt{nm} \log n)$ quantum queries by DFS (Lemma 5.1.2). The second step can be done with $O(\sqrt{nm})$ quantum queries by using minima finding (Theorem 3.1.10). □

By a more detailed analysis of the algorithm, it can be shown that the quantum query complexity of the strong graph connectivity problem is $O(n^{1.5})$ in the adjacency matrix model and $O(\sqrt{nm} \log n)$ in the adjacency list model. The quantum query lower bound can be proved with Theorem 4.1.3.

Theorem 5.2.6 [DHHM04] *The strong graph connectivity problem requires $\Omega(n^{1.5})$ quantum queries to the adjacency matrix and $\Omega(\sqrt{nm})$ quantum queries to the adjacency list.*

5.3 Minimum Spanning Tree and Shortest Paths

In this Section we present the quantum algorithms by Dürr et al. [DHHM04] for the minimum spanning tree and the shortest paths problem. These quantum algorithms use the MINIMUM TYPE FINDING quantum algorithm of Section 3.1 for finding the smallest values of different types.

Minimum Spanning Tree: Given a weighted undirected graph G, compute a spanning tree in G with minimal total edge weight.

Let $G = (V, E)$ be a directed graph with vertex set $V = \{v_1, \ldots, v_n\}$. The minimum spanning tree algorithm starts with n trees $T_1 = \{v_1\}, \ldots, T_n = \{v_n\}$. In each iteration of the algorithm, we search a minimum weight edge out of each tree, and add the edges to the trees. The algorithm stops, if there is only one tree left, which is a minimum spanning tree. The correctness of the algorithm follows from the fact that if U is a set of vertices of a connected graph G and $e \in (U \times (V \setminus U)) \cap E$ is a minimum weight edge, then there is a minimum spanning tree containing e.

We define two function f and g for using the MINIMUM TYPE FIND-ING algorithm. We replace every undirected edge $\{u, v\} \in E$ by two directed edges (u, v) and (v, u), and enumerate the edges from 1 to $2m$. We denote with:

- $f : [2m] \rightarrow \mathbb{N} \cup \infty$ the function that maps every directed edge (u, v) to its weight, if u and v belong to different trees of the current spanning forest, and to ∞ otherwise.

- $g : [2m] \rightarrow [d]$ the function that maps every directed edge (u, v) to the index j of the tree T_j containing u, where d is the current number of trees.

Now we can summarize the quantum algorithm:

Algorithm 7 MINIMUM SPANNING TREE
___
Input: Undirected weighted graph $G = (V, E)$.

Output: Minimum spanning tree $T = (V, A)$.

Complexity: **M:** $O(n^{3/2})$, **L:** $O(\sqrt{nm})$ quantum queries.

1: $d := n$, $A := \emptyset$

2: **while** $d > 1$ **do**

3:    $E_d :=$ MINIMUM TYPE FINDING$[f, g, d]$

4:    $A := A \cup E_d$

5:    $d := c(T)$

6:    Recompute f, g

7: $T = (V, A)$
___

Theorem 5.3.1 [DHHM04] *The quantum query complexity of the* MIN-IMUM SPANNING TREE *algorithm is* $O(n^{1.5})$ *in the adjacency matrix model and* $O(\sqrt{nm})$ *in the adjacency list model.*

Proof. The MINIMUM TYPE FINDING algorithm use $O(\sqrt{dm})$ quantum queries. At the beginning of the l'th iteration, the number of trees is $d = n/2^{l-1}$, and thus we use at most $O(\sqrt{nm/2^{l-1}})$ quantum queries to the adjacency list. In total the number of queries is at most

$$\sum_{l \geq 1} O\left(\sqrt{\frac{nm}{2^{l-1}}} \right) = O(\sqrt{nm}).$$

The matrix model is an instance of the list model with $m = n(n-1)$ edges. The overall error probability is bounded by $1/4$. $\qquad\square$

Now we consider the problem of finding all shortest paths from a vertex.

Shortest Paths: Given a weighted graph G with edge weight $c : E \to \mathbb{R}$

and a source vertex v_s, compute a tree T such that the shortest paths from v_s to all the other vertices is in T.

Classically the shortest path problem can be solved by Dijkstra's algorithm. The quantum algorithm by Dürr et al. [DHHM04] constructs a tree T, such that for every vertex v of G, the shortest path from v_s to v is in T. We denote with $d(v_s, v)$ the shortest path length from v_s to v. We called an edge $(u, v) \in E$ *border edge* of T, if $u \in T$ (*source vertex*) and $v \notin T$ (*target vertex*). Then it holds $d(v_s, v) = d(v_s, u) + c(u, v)$. The Dijkstra's algorithm starts with $T = \{v_s\}$ and iteratively adds the cheapest border edge to it.

Let P_l be a set of vertices of G, we compute $|P_l|$ cheapest border edges with disjoint target vertices. For this task we using the MINIMUM TYPE FINDING quantum algorithm. Let $N := \sum_{v \in P_l} d_G(v)$, and number all edges with source in P_l from 1 to N. We denote with:

- $g : [N] \to V$ the function, where $g(i)$ is the target vertex of the i'th edge.

- $f : [N] \to \mathbb{R}^+$ the function, where $f(i) := c(e_i)$ if $g(i) \notin T$, and ∞ otherwise.

Algorithm 8 SHORTEST PATHS

Input: Directed graph $G = (V, E)$, edge weight $c : E \to \mathbb{R}^+$, root $v_s \in V$.

Output: Shortest path tree T with root v_s.

Complexity: **M:** $O(\sqrt{nm} \log^2 n)$, **L:** $O(n^{1.5} \log^2 n)$.

1: $T := (\{v_s\}, \emptyset), l := 1, P_1 := \{v_s\}$

2: **while** $V(T) \neq V(G)$ **do**

3:　　$d := |P_l|$

4:　　$A_l :=$ MINIMUM TYPE FINDING$[f, g, d]$

5:　　Choose $(u, v) \in E$: $c((u, v)) = \min\{c((x, y)) \mid (x, y) \in \bigcup_i A_i, \ y \notin \bigcup_i P_i\}$

6:　　$V(T) := V(T) \cup \{v\}$

7:　　$E(T) := E(T) \cup \{(u, v)\}$

8:　　$P_{l+1} := \{v\}$

9:　　$l := l + 1$

10:　　**if** $l \geq 2 \wedge |P_{l-1}| = |P_l|$ **then**

11:　　　　$P_{l-1} := P_{l-1} \cup P_l$

12:　　　　$l := l - 1$

Theorem 5.3.2 [DHHM04] *The quantum query complexity of the* SHORTEST PATH *algorithm is* $O(\sqrt{nm} \log^2 n)$ *in the adjacency list model and* $O(n^{1.5} \log^2 n)$ *in the adjacency matrix model.*

By using a reduction from minima finding, it is simple to show a lower bound:

Theorem 5.3.3 [DHHM04] *The minimum spanning tree and the shortest path problem requires* $\Omega(n^{1.5})$ *quantum queries to the adjacency matrix and* $\Omega(\sqrt{nm})$ *quantum queries to the adjacency list.*

Remark 5.3.4 The quantum time complexity of the algorithms in this Chapter are by a $\log n$ factor bigger than its query complexity.

5.4 Maximum Flow

In this Section we present a quantum maximum flow algorithm by Ambainis and Špalek [AS06].

Definition 5.4.1 A *(flow) network* $N = (G, s, t, c)$ is a directed graph $G = (V, E)$ which the *capacity* function $c : V \times V \to \mathbb{R}^+$ and two distinct vertices $s, t \in V$ which are called the *source* and the *sink* vertex. If $(u, v) \notin E$, then $c(u, v) = 0$. A *flow* on N is a function $f : V \times V \to \mathbb{R}$ that satisfies the following three properties:

1. *Capacity constraint*: For all $u, v \in V : f(u, v) \leq c(u, v)$.

2. *Skew symmetry*: For all $u, v \in V : f(u, v) = -f(v, u)$.

3. *Flow conservation*: For all $u \in V \setminus \{s, t\} : \sum_{v \in V} f(u, v) = 0$.

The value $f(u, v)$ called the *flow* from vertex u to vertex v. The *size* $|f|$ of the flow f is defined as $|f| = \sum_{v \in V} f(s, v)$.

Here we consider the following maximum flow problem:

Maximum Flow: Given an integer flow network N with capacities bounded by U, compute a flow of N with maximum size.

Definition 5.4.2 Let $N = (G, s, t, c)$ be a network:

- Given a flow f on N, the *residual capacity* for f is $r(u, v) := c(u, v) - f(u, v)$ for each pair of vertices $u, v \in V(G)$. The *residual graph* for the flow f is $R = (V, E_f)$ with

$E_f = \{(u, v) \mid r(u, v) > 0\}$. A *residual flow* is the difference between an optimal flow f^* and the current flow f: if $f^*(a) \geq f(a)$ then the residual flow on a is $f^*(a) - f(a)$, and otherwise the residual flow is $f(a) - f^*(a)$.

- An *augmenting path* in a network is a path p from the source to the sink whose residual capacity is bigger than zero. The *bottleneck capacity* of p is the minimum residual capacity along p.

- A *blocking flow* in a subgraph H of G containing s and t, is a flow where every directed s-t-path contains an arc with zero residual capacity.

- The *level* of a vertex v, denoted by level(v), is the least number of edges in a path from s to v. The *layered network* of N is the directed subgraph $L = (V, E')$ of G, where $E' = \{(u, v) \in E \mid \text{level}(v) = \text{level}(v) + 1\}$.

The layered network contains all shortest length augmenting paths from the source to the sink, which are of the same length. The construction of the layered network can be done with quantum breadth first search.

Lemma 5.4.3 [AS06] *There is a quantum algorithm that computes a layered network in time $O(n^{3/2} \log^2 n)$ in the adjacency matrix model and $O(\sqrt{nm} \log^2 n)$ in the adjacency list model.*

Lemma 5.4.4 [AS06] *There is a quantum algorithm for computing a blocking flow in the layered network of depth j with running time of $O((\sqrt{jmA_jU} + \sqrt{nm})\log^2 n)$ in the adjacency list model, where A_j is the size of its blocking flow.*

Proof. The classical algorithm for determining a blocking flow in the layered network of depth j is the following:

1. Compute the layered network L of G, and mark all vertices as enabled.

2. While the bottleneck capacity Δ_p is greater than zero:

   (a) DFS a path p over enabled vertices in L from s to t.

   (b) Disable all vertices from which there is no path to t.

   (c) Augment f by Δ_p along p.

   (d) Update L along p.

Now we analyse the running time complexity of this procedure in the quantum case. Let a_v be the number of augmenting paths going through vertex v. Finding the edges of the augmenting paths during v can be done with quantum search in time

$$\sum_{i=1}^{a_v} \sqrt{\frac{U d_G(v)}{i}} = O(\sqrt{U d_G(v) a_v}).$$

It holds $\sum_v a_v \leq j A_j$, then the total time for finding a blocking flow is

$$\sum_v \sqrt{U d_G(v) a_v} \leq \sqrt{U} \sqrt{\sum_v d_G(v)} \cdot \sqrt{\sum_v a_v} = O(\sqrt{U m j A_j}).$$

93

Disabling all vertices from which there is no path to t can be done by the quantum BFS procedure in $O(\sqrt{nm})$ steps. $\qquad\square$

Now we are able to present the quantum algorithm by [AS06] for the maximal flow problem, by using the following simple fact from flow theory:

Lemma 5.4.5 [ET75] *Given an integer flow network with capacities bounded by U, whose layered residual network has depth j, then the size of the residual flow is at most $\min((2n/j)^2, m/j) \cdot U$.*

Theorem 5.4.6 [AS06] *Let $U \le n^{1/4}$. There is a quantum algorithm for the maximum flow problem with running time of $O(n^{13/6} \cdot U^{1/3} \log^4 n)$ in the adjacency matrix model and $O(\min\{n^{7/6}\sqrt{m} \cdot U^{1/3}, m\sqrt{nU}\} \log^4 n)$ in the adjacency list model.*

Proof. The algorithm consists of the following two parts (analog to [ET75]):

1. Iteratively augment the current flow by blocking flows in the layered residual networks, until the depth of the network is $k = \min(n^{2/3}U^{1/3}, \sqrt{mU})$.

2. Search single augmenting paths, while there are some.

We determine the quantum running time complexity of this algorithm. By using Lemma 5.4.4, the first part can be done in

$$\sqrt{mU} \cdot \sum_{j=1}^{k} \sqrt{jA_j} + k\sqrt{nm}.$$

After the first part, the algorithm has constructed a layered network of depth k. By Lemma 5.4.5, the residual flow has size

$$O(\min((n/k)^2, m/k) \cdot U) = O(k).$$

Therefore the search for augmenting paths terminates in $O(k)$ more iterations, and part two can be done in time $O(k\sqrt{nm})$ by using the quantum DFS procedure. Then the total running time of the maximum flow algorithm is

$$O\left(\sqrt{mU} \cdot \sum_{j=1}^{k} \sqrt{jA_j} + k\sqrt{nm}\right).$$

Now we prove that $\sum_j \sqrt{jA_j} = O(k^{3/2} \log^2 k)$. By Lemma 5.4.5, the residual flow after $l = k/2^i$ iterations is at most

$$O(\min((n/k)^2 \cdot 2^{2i}, m/k \cdot 2^i) \cdot U) \leq O(2^{2i}k) = O(k^3/l^2).$$

Then it follows that $\sum_{j=l}^{2l-1} A_j = O(k^3/l^2)$, and we get

$$\sum_{j=1}^{k} \sqrt{jA_j} = \sum_{i=0}^{\log k} \sum_{j=2^i}^{2^{i+1}-1} \sqrt{jA_j} \leq \sum_{i=0}^{\log k} \sqrt{\sum_{j=2^i}^{2^{i+1}-1} j} \sqrt{\sum_{j=2^i}^{2^{i+1}-1} A_j}$$

$$\leq \sum_{i=0}^{\log k} \sqrt{2^i \cdot 2^{i+1}} \sqrt{\sum_{j=2^i}^{2^{i+1}-1} A_j} \leq \sqrt{2} \sum_{i=0}^{\log k} 2^i \sqrt{\frac{k^3}{2^{2i}}}$$

$$= O(k^{3/2} \log^2 k).$$

Since $U \leq n^{1/4}$ then $kU \leq n$. Then the total running time is

$$O(k\sqrt{m}(\sqrt{kU} \log^2 k + \sqrt{n})) = O(k\sqrt{nm} \log^2 k)$$
$$= O(\min(n^{7/6}\sqrt{m} \cdot U^{1/3}, \sqrt{nU}m) \log^2 n).$$

The time complexity for the adjacency model follows from setting $m = n^2$. In order to get the success probability of $1 - 1/n$, we need to amplify the success probability of each subroutine by repeating it $O(\log n)$ times, see Remark 3.1.9. Considering this, we obtain the indicated quantum time complexity. □

Remark 5.4.7 If $m = \Omega(n^{1+\varepsilon})$ for some $\varepsilon > 0$ and U is small, then the algorithm is polynomially faster than the best known classical algorithm. For constant U and $m = O(n)$, it is slower by at most a log-factor.

Chapter 6

Matching Problems

In this Chapter we present quantum algorithms for matching problems. A matching in a graph is a set of edges such that for every vertex at most one edge of the matching is incident on the vertex. The task is to find a matching of maximum cardinality. The matching problem has many important applications in graph theory and computer science. We study the complexity of algorithms for matching problems on quantum computers and compare these to the best known classical algorithms. We will consider different versions of the matching problems, depending on whether the graph is bipartite or not and whether the graph is unweighted or weighted. The results of this Chapter are published in [Doe08].

In Section 6.1 we present quantum algorithms for unweighted matching problems. First we give a quantum algorithm for computing a maximal matching in general graphs in time $O(\sqrt{nm}\log^2 n)$. Then we consider the maximum matching problem. The best classical algorithms for finding a matching of maximum cardinality based on augmenting paths with running time $O(\sqrt{nm})$ (Micali and Vazirani [MV80]) and on matrix multiplication with running time $O(n^\omega)$ (Mucha and Sankowski [MS04]) where $\omega \leq 2.38$. Ambainis and Špalek [AS06] published an $O(n^2(\sqrt{m/n} + \log n) \log^2 n)$ quantum time algorithm for computing a maximum matching. But this quantum algorithm is in no case better than the best classical algorithm from [MV80] and [MS04]. The authors of [AS06] state as an open question, to improve with a quantum algorithm the fastest known classical algorithm by Micali and Vazirani [MV80]. We solve this question by presenting an $O(n\sqrt{m} \log^2 n)$ quantum time algorithm for finding a maximum matching in general graphs. Our algorithm is faster than the best classical algorithms for graphs with $m > n \log^4 n$.

In Section 6.2 we give quantum algorithms for weighted matching problems. The best classical algorithms for computing a maximum weight matching in bipartite graphs were developed by Gabow and Tarjan [GT89] with running time $O(\sqrt{n}m \log(nN))$ and Kao et al. [KLST01] with time complexity $O(\sqrt{n}W/k(n, W/N))$, where $k(x, y) = \log x / \log(x^2/y)$, N is the largest edge weight and W is the total edge weight of G. We construct a quantum algorithm using the decomposition theorem of [KLST01] with running time of $O(n\sqrt{m}N \log^2 n)$ for maximum weight matching in bipartite graphs.

6.1 Unweighted Matchings

Definition 6.1.1 Let $G = (V, E)$ be an undirected graph, a *matching* is a subset $M \subseteq E$ such that for all vertices $v \in V$, at most one edge of M is incident on v. We say that a vertex $v \in V$ is *matched* by M if some edge in M is incident on v, otherwise v is called *unmatched* or *free*. A matching is called *perfect*, if all vertices of G are matched. The set M is called *maximal*, if there is no matching $M' \subseteq E$ with $M \subset M'$. A *maximum matching* is a matching of maximum cardinality. For a matching M, a path P in G is called *alternating* in M if edges in P are alternately in M and not in M. An alternating path P is called *augmenting path* for the matching M if the two end vertices of P are unmatched by M.

We regard the following matching problems in unweighted graphs:

Maximal Matching: Given a graph G, compute a maximal matching.

Maximum Matching: Given a graph G, compute a maximum matching.

Maximal Matching

In this subsection we construct a quantum query algorithm for the maximal matching problem. Then we prove that this algorithm is nearly optimal in the adjacency matrix model.

Theorem 6.1.2 *The quantum query complexity of the maximal matching problem is $O(n^{1.5} \log n)$ in the adjacency matrix model and*

$O(\sqrt{nm}\log n)$ *in the adjacency list model.*

Proof. Let $G = (V, E)$ be a graph with vertex set $V = \{v_1, \ldots, v_n\}$ and $k = 1$ an integer. We compute a set M of matching edges which is maximal in G. At the beginning $M = \emptyset$ and we mark every vertex of G as enabled. While there are some enabled vertices of G, we search with the Grover algorithm an adjacent edge of the vertex v_k in G. If there is no such edge, we mark the vertex v_k as disabled. Otherwise we use the edge $\{v_k, v\}$ for the matching M, and we mark the two vertices v_k and v as disabled. Then we increase the value of k at one.

In the matrix model, one search can be done in $O(\sqrt{n})$ quantum queries. In total we use $O(n^{1.5})$ quantum queries to the adjacency matrix for computing a maximal matching. In the list model $\sqrt{d_G(v_k)}$ queries are required for every vertex v_k, and in total we use $\sum_{k=1}^{n}\sqrt{d_G(v_k)} = O(\sqrt{nm})$ quantum queries. In order to get a constant success probability, we need to amplify the success probability of each subroutine by repeating it $O(\log n)$ times, see Remark 3.1.9. Then we obtain a maximal matching M in the indicated quantum query complexity. $\square$

Corollary 6.1.3 *The quantum time complexity of the maximal matching problem is $O(n^{1.5}\log^2 n)$ in the adjacency model and $O(\sqrt{nm}\log^2 n)$ in the adjacency list model.*

Berzina et al. [BDFLS04] and Zhang [Zha04] showed that $\Omega(n^{3/2})$ quantum queries to the adjacency matrix are required for computing a maximum matching in a bipartite graph. This implies a quantum lower bound of $\Omega(n^{3/2})$ for the quantum time complexity. Since the bipartite maximum matching problem is a special case of the other maximum matching problems, it follows that the $\Omega(n^{3/2})$ lower bound holds for all the maximum matching problems considered here. We show that the same lower bound holds also for the maximal matching problem. Therefore we have determined the exact quantum query complexity of the maximal matching problem.

Theorem 6.1.4 *The maximal matching problem requires $\Omega(n^{1.5})$ quantum queries to the adjacency matrix.*

Proof. We construct the sets A and B for the usage of Theorem 4.1.1. Let f be the Boolean function which is one, iff there is a maximal matching set of size n. The set A consists of all graphs $G = (V, E)$ with $|V| = 3n + 1$ satisfying the following requirements: 1. There are n mutually not connected red vertices. 2. There are $2n$ green vertices not connected with the red ones. Green vertices are grouped in pairs and each pair is connected by edge. 3. There is a black vertex which is connected to all red and green vertices. Let M be the set of edges between every pair of green vertices of G. Then M is a maximal matching in G of size n. The value of the function f for all graphs $G \in A$ is 1.

The set B consists of all graphs $G' = (V, E)$ with $|V| = 3n + 1$ satisfying the following requirements: 1. There are $n - 2$ mutually not connected red vertices. 2. There are $2n + 2$ green vertices not connected with red ones, green vertices are grouped in pairs and each pair is connected by edge. 3. There is a black vertex which is connected to all red and green vertices. The value of the function f for all graphs $G' \in B$ is 0, since there no maximal matching of size n in G'.

From each graph $G \in A$, we can obtain $G' \in B$ by adding one edge between two red vertices (then the two vertices become green), then $l = \Omega(n^2)$. From each graph $G' \in B$, we can obtain $G \in A$ by deleting an edge between two green vertices, then $l' = \Omega(n)$. By Theorem 4.1.1, the quantum query complexity is $\Omega(\sqrt{l \cdot l'}) = \Omega(n^{1.5})$. $\qquad\qquad\qquad\square$

Maximum Matching

The best classical algorithm for finding a matching of maximum cardinality in general graphs is based on augmenting paths and has running time $O(\sqrt{n}m)$, see Micali and Vazirani [MV80]. Mucha and Sankowski [MS04] presented an algorithm based on matrix multiplication that finds a maximal matching in general graphs in time $O(n^\omega)$, where $2 \leq \omega \leq 2.38$ is the exponent of the best matrix multiplication algorithm.

Ambainis and Špalek [AS06] constructed a $O(n^2(\sqrt{m/n}+\log n)\log^2 n)$ quantum algorithm for computing a maximum matching. Unfortunately this quantum algorithm is in no case better than the best classical algorithm from [MV80] and [MS04].

In this subsection we present an $O(n\sqrt{m}\log^2 n)$ quantum time algorithm for finding a maximum matching in a general graphs. This algorithm is faster than the best classical one for computing a maximum matching in a graph with $m > n \log^4 n$. For bipartite graphs the best classical matching algorithm has running time $O(\sqrt{n}m)$ (see Hopcroft and Karp [HK73]). Ambainis and Špalek [AS06] improved this bound with an $O(n\sqrt{m}\log^2 n)$ quantum algorithm for computing a maximum matching in these graphs. The algorithm is polynomially faster than the best classical algorithm, but not necessarily optimal. The time complexity of our quantum algorithm for the maximum matching problem in general graphs matches the complexity given in the algorithm from [AS06] for the restricted case of bipartite graphs.

We speed up the $O(\sqrt{n}m)$ algorithm by Micali and Vizirani [MV80]. This algorithm works in phases. In each phase a maximal set of disjoint minimum length augmenting paths is found, and the existing matching is increased along this paths. We implement such a phase in time $O(\sqrt{mn}\log^2 n)$ with quantum search. Only $O(\sqrt{n})$ such phases are needed for finding a maximum matching, see [HK73], [MV80]. Before we explain the maximum matching algorithm, we give some important definitions.

Definition 6.1.5 Let $G = (V, E)$ be a graph and M be a matching in G.

- The *evenlevel* (*oddlevel*) of a vertex v is the length of a minimum even (odd) length alternating path from v to a free vertex, if there is one, and infinite otherwise.

- The *level* is the length of a minimum alternating path from v to a

103

free vertex.

- A vertex is *outer* (*inner*) iff its level is even (odd). If v is outer (inner) then its oddlevel (evenlevel) will be refered to as the *other level* of v.

- An edge (u, v) in a graph G with matching M is called a *bridge* if either both evenlevel(u) and evenlevel(v) are finite, or both oddlevel(u) and oddlevel(v) are finite.

- A *blossom* B in a matched graph G is a cycle of odd length k, in which the edges are maximally matched. In every blossom there is one free vertex, called *basis* of B. The two vertices at distance $\lfloor k/2 \rfloor$ from the basis are called *peaks* of B.

- Let u be a vertex of G which is not matched. If u is inner and oddlevel(u) $= 2i + 1$ then v is called a *predecessor* of u iff evenlevel(v) $= 2i$ and $(u, v) \in E$. If u is outer then v is a predecessor of u iff (u, v) is a matched edge.

- The *ancestor* is the transitive closure of the relation predecessor.

The classical algorithm by [MV80] for computing a maximal matching in a general graph consists of a main routine SEARCH, and three subroutines: BLOSS-AUG, FINDPATH and TOPOLOGICAL ERASE. We describe shortly the four parts of the algorithm, for details see [MV80]:

SEARCH:

Given a graph $G = (V, E)$ and a matching M, SEARCH constructs simultaneously for every free vertex v of G a Breadth First Search (BFS) tree which is rooted at v, to find the oddlevel and evenlevel of each vertex in G. At the start of the subroutine the two levels of each vertex of G are set to infinity. Then SEARCH grows the BFS trees by incrementing the search level by one each time.

When SEARCH detects that a certain edge (u, v) is a bridge, it calls the subroutine BLOSS-AUG with the parameter u and v.

BLOSS-AUG:

This subroutine is called with vertices u and v such that the edge (u, v) is a bridge. The result is either the formation of a new blossom, or a minimum augmenting path. A new blossom is formed if and only if the following condition holds:

1. There exist a vertex z such that z is an ancestor of vertex u and v.

2. The vertices u and v do not have any ancestors, other than z, whose level is equal to the level of z.

If the condition holds for bridge (u, v) and b is the vertex whose level is maximum, then the new blossom is the set of vertices w such that:

1. Vertex w does not belong to any other blossom when B is formed.

2. Either $w = u$ or $w = v$ or w is an ancestor of u or w is an ancestor of v.

3. The vertex b is an ancestor of w.

From this follows that b is the base of B and u and v are the peaks of B.

BLOSS-AUG performs a Double Depth First Search (DDFS), consisting in growing two DFS trees T_1 and T_r simultaneously, i.e. if at a certain stage, the centers of activities of T_1 and T_r are at v_1 and v_r, then the DDFS grows T_1 if level$(v_1) \geq$ level(v_r), and its grows T_r otherwise. T_1 and T_r are rooted at u and v.

For the special details of the DDFS see [MV80, page 21]. During the DDFS, the two trees can find two different free vertices, then an augmentation of the matching is possible.

FINDPATH:

When BLOSS-AUG finds the presence of a minimum augmenting path, we use FINDPATH to search for such a path P. FINDPATH is called with two vertices v_h and v_l and a blossom B as parameters. It holds level$(v_h) \geq$ level(v_l) and they both belong to a common minimum augmenting path.

The procedure returns a path between v_l and v_r with a DFS starting at v_h to find v_l. The present matching is increased along the minimum augmenting path P.

TOPOLOGICAL ERASE:

After FINDPATH has found the minimum augmenting path P and the matching has been increased along this path, this subroutine deletes from the graph the path P and all those edges which cannot be part of a minimum augmenting path disjoint from P.

This subroutine uses a topological sort. Each vertex has a counter which at any stage indicates the number of its unerased predecessor edges. A vertex is deleted with all incident edges, either when its counter is zero or when it enters a minimum augmenting path detected by FINDPATH. The counter of the free vertices is one at the start and during the phase, since they have no predecessor.

Theorem 6.1.6 *There is a quantum algorithm for the maximum matching problem with running time of $O(n^2 \log^2 n)$ in the adjacency matrix model and $O(n\sqrt{m} \log^2 n)$ in the adjacency list model.*

Proof. We show that a phase consisting of the four subroutines SEARCH, BLOSS-AUG, FINDPATH and TOPOLOGICAL ERASE can be implemented with quantum search in time $O(n^{1.5} \log^2 n)$ in the adjacency matrix model and in time $O(\sqrt{nm} \log^2 n)$ in the adjacency list model. Only $O(\sqrt{n})$ such phases are needed for finding a maximum matching, see [HK73], [MV80]. We regard the above four subroutines.

In the SEARCH procedure we perform a Breadth First Search to find the evenlevel and oddlevel of each vertex in G as follows: All free vertices of G get the evenlevel 0 and the other levels are infinite. SEARCH constructs simultaneously for every free vertex v of G a BFS tree to find the two level numbers of each vertex in G. In the adjacency matrix model a vertex is found in $O(\sqrt{n})$ quantum queries. Every vertex is processed at most twice, since we have two level numbers for each vertex. In the list model, processing a vertex v costs $O(\sqrt{d_G(v) \cdot n_v})$ quantum queries, where n_v is the number of vertices adjacent to v with a infinite level number. Since $\sum_v n_v \leq 2n$, then the total quantum query complexity is upper-bounded by the Cauchy-Schwarz inequality:

$$\sum_v \sqrt{d_G(v) n_v} \leq \sqrt{\sum_v d_G(v)} \sqrt{\sum_v n_v} = O(\sqrt{mn}).$$

The procedure BLOSS-AUG performs a Double Depth First Search, consisting of growing two DFS trees. With quantum search we perform these two DFS in $O(n^{1.5})$ quantum queries to the adjacency matrix model and in $O(\sqrt{nm})$ quantum queries to the adjacency list model, see Lemma 5.1.2.

The subroutine FINDPATH returns a path with a DFS starting at v_h to find v_l. Clearly with Lemma 5.1.2, the quantum query complexity is $O(n^{1.5})$ in the adjacency matrix model and $O(\sqrt{nm})$ in the adjacency list model.

The quantum query complexity of the TOPOLOGICAL ERASE procedure is the same as the above three subroutines, by using Lemma 5.1.4

In order to get a constant success probability, we need to amplify the success probability of each subroutine by repeating it $O(\log n)$ times.

□

6.2 Weighted Matchings

In this Section, we look at weighted matchings in bipartite graphs. Let $G = (V, E)$ be a graph, $w(u, v)$ is the weight of an edge $\{u, v\} \in E$, if u is not adjacent to v, let $w(u, v) = 0$. We denote by N the largest edge weight and with W the total edge weight of G.

We regard the following two matching problems in weighted graphs:

Maximum Weight Bipartite Matching: Given a bipartite graph G with positive integer weights on the edges and without isolated vertices, compute a matching M in G such that the sum of the weights of the edges in M is maximum over all possible matchings.

Minimum Weight Perfect Bipartite Matching: Given a weighted bipartite graph G, compute a perfect matching M in G such that the sum of the weights of the edges in M is minimum over all possible perfect matchings.

Maximum Weight Bipartite Matching

The best classical algorithms for computing a maximum weight matching in bipartite graphs with positive integer weights have been developed by Gabow and Tarjan [GT89] with running time $O(\sqrt{n}m \log(nN))$ and by Kao, et al. [KLST01] with time complexity $O(\sqrt{n}W/k(n, W/N))$, where $k(x, y) = \log x / \log(x^2/y)$.

We give a quantum algorithm with running time $O(n\sqrt{m}N \log^2 n)$

for computing a maximum weight matching in bipartite graphs. If the largest edge weight N is constant, then we get an $O(n\sqrt{m}\log^2 n)$ time algorithm. For the construction of our quantum algorithm we use the decomposition theorem of [KLST01]. First we need some definitions and facts about a minimum weight cover in a bipartite graph.

Definition 6.2.1 Let $G = (X \cup Y, E)$ be a bipartite graph. A *cover* of G is a function $C : X \cup Y \to \mathbb{N}$ such that $C(x) + C(y) \geq w(x, y)$ for all $x \in X$ and $y \in Y$. Let $w(C) := \sum_{z \in X \cup Y} C(z)$ be the weight of C. The cover C is of *minimum weight*, if $w(C)$ is the smallest possible value.

The algorithms for computing a maximum weight matching in bipartite graphs use the following problem:

Minimum Weight Cover: Given a bipartite graph G with positive integer weights, compute a minimum weight cover of G.

Remark 6.2.2 A minimum weight cover is dual to the maximum weight matching in a bipartite graph G, i.e. from a maximum matching in G we can find a minimum weight cover of G in time $O(m)$, see [BM76].

Definition 6.2.3 Let $G = (X \cup Y, E)$ be a bipartite graph and $h \in \{1, \ldots, N\}$ be an integer. Define G_h as the graph which is formed by the edges $\{u, v\}$ of G with $w(u, v) \in [N - h + 1, N]$ and the edge weight $\{u, v\}$ of G_h is $w(u, v) - (N - h)$.

Let C_h be a minimum weight cover of G_h and G_h^* is formed by the edges $\{u, v\}$ of G with $w(u, v) - C_h(u) - C_h(v) > 0$ and the edge weight

$\{u, v\}$ of G_h^* is $w(u, v) - C_h(u) - C_h(v)$.

Now we present the Decomposition theorem of [KLST01].

Theorem 6.2.4 [KLST01] *Let h, G, G_h, C_h, G_h^* as in Definition 6.2.3 and let C_h^* be any minimum weight cover of G_h^*. If $D : X \cup Y \rightarrow \mathbb{N}$ is a function such that for every $u \in V(G)$,*

$$D(u) = C_h(u) + C_h^*(u),$$

then D is a minimum weight cover of G.

Using this theorem, a minimum weight cover of G can be computed with the following recursive procedure, see [KLST01].

Algorithm 9 MINIMUM WEIGHT COVER

Input: Bipartite graph $G = (X \cup Y, E)$ with positive integer weights.

Output: Minimum weight cover $D : X \cup Y \rightarrow \mathbb{N}$.

Complexity: **M:** $O(n^2 N \log^2 n)$, **L:** $O(n\sqrt{m}N \log^2 n)$ quantum steps.

1: Construct G_1 from G.

2: Find a minimum weight cover C_1 of G_1.

3: Construct G_1^* from G and C_1.

4: **if** $G_1^* = \emptyset$ **then**

5:    **return**$[C_1]$

6: **else**

7:    $C_1^* :=$ MINIMUM WEIGHT COVER$[G_1^*]$

8:    $D(u) := C_1(u) + C_1^*(u)$ for all u in G

9:    **return**$[D]$

The correctness of the algorithm follows from Theorem 6.2.4. We use our maximum matching quantum algorithm for computing a minimum weight cover of the graph G.

Theorem 6.2.5 *The quantum time complexity of the* MINIMUM WEIGHT COVER *algorithm is* $O(n\sqrt{m}N\log^2 n)$ *in the adjacency list model and* $O(n^2 N \log^2 n)$ *in the adjacency matrix model.*

Proof. We analyse the running time in the adjacency list model. We initialize a maximum heap in $O(m)$ time to store the edges of G according to their weights. Let $T(n, W, N)$ be the running time of the MINIMUM WEIGHT COVER quantum algorithm. Let L be the set of all edges in G_1, i.e. the heaviest edges in G. Then Step 1 takes $O(|L|\log m)$ time. In Step 2, we can compute a maximum matching of G_1 in $O(n\sqrt{|L|}\log^2 n)$ time steps, by using the maximum matching quantum algorithm. From this matching, C_1 can be found in $O(|L|)$ time, see [BM76]. Let L_1 be the set of the edges of G adjacent to some vertex u with $C_1(u) > 0$. Step 3 updates every edge of L_1 in $O(l_1 \log m)$ time, where $l_1 = |L_1|$.

The total running time of steps 1 to 3 is $O(n\sqrt{l_1}\log^2 n)$, since $L \subseteq L_1$. The total weight of G_1^* is at most $W - l_1$. Step 7 uses then at most $T(n, W - l_1, N')$ time, where $N' < N$ is the maximum edge weight of G_1^* and it follows

$$T(n, W, N) = O(n\sqrt{l_1}\log^2 n) + T(n, W - l_1, N'),$$

where $T(n, 0, N') = 0$. We apply the procedure recursively for some positive integers $l_1, l_2, \ldots, l_p$ with $p \leq N$ and $\sum_{1 \leq i \leq p} l_i = W$, it follows

$$T(n, W, N) = O\left(n\log^2 n \sum_{i=1}^{p} \sqrt{l_i}\right).$$

Since $\sum_{i=1}^{p} \sqrt{l_i} \leq \sqrt{p \sum_{i=1}^{p} l_i}$, then

$$T(n, W, N) = O\left(n \log^2 n \sqrt{p \sum_{i=1}^{p} l_i}\right) = O\left(n \log^2 n \sqrt{NW}\right).$$

Since $W \leq Nm$ it follows $T(n, W, N) = O(n\sqrt{m}N \log^2 n)$ in the list model. The running time for the matrix model follows setting $m = n^2$.

$\square$

Now we use the algorithm by [KLST01] to recover a maximum weight matching of a bipartite graph G from a minimum weight cover of G.

Algorithm 10 MAXIMUM WEIGHT MATCHING

Input: Bipartite graph $G = (X \cup Y, E)$ with positive integer weights.

Output: Maximum weight matching M.

Complexity: **M:** $O(n^2 N \log^2 n)$, **L:** $O(n\sqrt{m}N \log^2 n)$ quantum steps.

1: $D :=$ MINIMUM WEIGHT COVER$[G]$
2: $A := \{\{u, v\} \in E \mid w(u, v) = D(u) + D(v)\}$
3: $H := (V, A)$
4: Make two copies of H, call them H^a and H^b. For each vertex u of H, let u^a and u^b denote the corresponding vertex in H^a and H^b.
5: Union H^a and H^b to form H^{ab}, and add to H^{ab} the set of edges
   $\{(u^a, u^b) \mid u \in V(H), \ D(u) = 0\}$.
6: Find a maximum matching K of H^{ab}.
7: $M = \{(u, v) \mid (u^a, v^a) \in K\}$

Theorem 6.2.6 *There is a quantum algorithm for computing a maximum weight matching in a bipartite graph with running time of $O(n\sqrt{m}N \log^2 n)$ in the adjacency list model and $O(n^2 N \log^2 n)$ in the adjacency matrix model.*

Proof. We use the above algorithm. The graph H^{ab} has at most $2n$ nodes and at most $3m$ edges. The maximum matching K can be constructed by a quantum algorithm in time $O(n\sqrt{m}\log^2 n)$ in the adjacency list model and in time $O(n^2 \log^2 n)$ in the adjacency matrix model. $\qquad\square$

Corollary 6.2.7 *There is a quantum algorithm for computing a maximum weight matching in a bipartite graph with constant edge weight with running time of $O(n\sqrt{m}\log^2 n)$ in the adjacency list model and $O(n^2 \log^2 n)$ in the adjacency matrix model.*

Minimum Weight Perfect Bipartite Matching

Now we give the quantum time complexity for computing a minimum weight perfect matching in a bipartite graph. The best classical algorithm for computing a minimum weight perfect matching in a bipartite graph computes shortest paths in a graph, see Cook et al. [CCPS98]. The time complexity for such an algorithm is given by the following Theorem.

Theorem 6.2.8 [see CCPS98] *There is an algorithm for the minimum weight perfect matching problem for bipartite graphs with running time $O(nS(n, m))$, where $S(n, m)$ is the time needed to solve the shortest path problem on a digraph with n vertices and m arcs.*

For the shortest path problem we can use the quantum algorithm by Dürr et al. [DHHM04], and it follows:

Theorem 6.2.9 *There is a quantum algorithm for the minimum weight perfect bipartite matching problem with running time of $O(n^{2.5} \log^2 n)$ in the adjacency matrix model and $O(n\sqrt{nm} \log^2 n)$ in the adjacency list model.*

Chapter 7

Graph Traversal Problems

In this Chapter we study the quantum complexity of algorithms for graph traversal problems. More precisely, we look at eulerian tours, hamiltonian tours, optimal postman tours, travelling salesman problem and project scheduling. The results of this Chapter are published in [Doe07b] and [Doe07c].

In Section 7.1 we study the quantum query complexity of the eulerian graph problem. In this problem we have to decide if a graph G has an eulerian cycle, this is a closed walk that contains every edge of G exactly once. We compute the precise quantum query complexity of the eulerian graph problem in the adjacency matrix and the list model.

In Section 7.2 we consider the optimal postman tour problem. We have given a weighted graph G, and compute a closed walk with minimum total edge weight that uses each edge of G at least once. There is a classical algorithm by Edmonds and Johnson [EJ73], which solves this problem in $O(n^3 + m)$ time steps. We improve this upper bound by a quantum algorithm with running time of $O(n\sqrt{nm}\log^2 n)$. Furthermore, we show an $\Omega(n^2)$ quantum query lower bound for the optimal postman tour problem.

In Section 7.3 we consider the hamiltonian cycle problem, an important **NP**-complete graph problem. A hamiltonian cycle of a graph G is a cycle which contains all the vertices of G exactly once. Berzina et al. [BDFLS04] proved that the hamiltonian cycle problem requires $\Omega(n^{1.5})$ quantum queries to the adjacency matrix. We show an $O(n^{2n/(n+1)})$ quantum query upper bound for this problem by using the quantum walk technique.

In Section 7.4 we study the travelling salesman problem in graphs with maximal degree three, four and five. Eppstein [Epp03] constructed algorithms for the travelling salesman problem on graphs with bounded degree three and four which are faster than $O(2^n)$. We show that with a quantum computer we can solve the travelling salesman problem on graphs with maximal degree three, four and five quadratically faster than in the classical case.

In Section 7.5 we consider a project scheduling problem. A digraph model can be used to schedule projects consisting of several interrelated tasks. Some of these tasks can be executed simultaneously, but some tasks cannot begin until certain others are completed. We present an optimal quantum query algorithm for computing the earliest completion time for every vertex of the network.

7.1 Eulerian Tour

In this Section we consider the following problem:

Eulerian Graph: Given a connected graph G, decide if G has a closed walk that contains every edge of G once.

It is a well known fact in graph theory that a graph G is eulerian iff the degree of every vertex in G is even.

Theorem 7.1.1 *The quantum query complexity of the eulerian graph problem is $O(\sqrt{n})$ in the adjacency list and $O(n^{1.5})$ in the adjacency matrix model.*

Proof. In the adjacency list model, the degree of every vertex is given. We search an odd number in the degree list. If there is a vertex in the graph G with odd degree, then G is not eulerian. This simple quantum search can be done in $O(\sqrt{n})$ quantum queries to the degree list.

In the adjacency matrix model, we search a vertex with odd degree (if there is one). We use Grover search in combination with a classical algorithms for computing the parity in every row in the adjacency matrix of G. The total quantum query complexity of the eulerian graph problem is $O(n^{1.5})$ in the adjacency matrix model. □

Corollary 7.1.2 *There is a quantum algorithm for the eulerian graph problem with time complexity of $O(\sqrt{n}\log n)$ in the adjacency list and $O(n^{1.5}\log n)$ in the adjacency matrix model.*

Now we show that our upper bounds are tight in the matrix and list models.

Theorem 7.1.3 *The eulerian graph problem requires $\Omega(\sqrt{n})$ quantum queries to the adjacency list and $\Omega(n^{1.5})$ quantum queries to the adjacency matrix.*

Proof. In the adjacency matrix model, we reduce OR of n parities of length n to the eulerian tour problem. We define

$$z := (x_{1,1} \oplus \ldots \oplus x_{1,n}) \vee \ldots \vee (x_{n,1} \oplus \ldots \oplus x_{n,n}).$$

It is a well known fact that the computation of z requires $\Omega(n^{1.5})$ quantum queries (see [Amb02]). Then it is $z = 0$ iff the graph G with adjacency matrix $A = (x_{i,j})$ has an eulerian tour. In the adjacency list model, the $\Omega(\sqrt{n})$ lower bound follows by a simple reduction from the Grover search. $\square$

Since the upper and the lower bounds match, we have determined the precise quantum query complexity of the eulerian graph problem.

7.2 Optimal Postman Tour

The chinese mathematician Guan [Gua62] introduced the following problem:

Optimal Postman Tour: Given a weighted graph G, compute a closed walk of minimum total edge weight that uses each edge at least once.

In the case that each vertex of the graph has even degree, then any eulerian tour is an optimal postman tour. Otherwise, some edges must be used more than once. Every postman tour in G corresponds to an eulerian tour of a graph G^* formed from G, by adding to G as many additional copies minus one of an edge as the number of times it was used during the postman tour.

We use the classical algorithm by Edmonds and Johnson [EJ73], and speed up this algorithm by the following two quantum graph subroutines (see Chapter 5.3 and 6.2):

- SHORTEST PATH$[G, u, c]$ is the quantum algorithm to compute a tree P_u, such that the shortest paths with respect to the edge weight function $c : E \rightarrow \mathbb{R}^+$ from u to all the other vertices of G are contained in the tree (we denote by $P_{u,v}$ the shortest path from u to v in P_u).

- MINIMUM WEIGHT PERFECT BIPARTITE MATCHING$[G, c]$ is the quantum algorithm to compute a minimum weight perfect matching in a bipartite graph G with edge weight function $c : E \rightarrow \mathbb{R}^+$.

The classical optimal postman algorithm by Edmonds and Johnson [EJ73] has running time of $O(n^3 + m)$. We improve the running time to $O(n\sqrt{nm}\log^3 n)$ by the following quantum algorithm:

121

Algorithm 11 OPTIMAL POSTMAN TOUR
___
Input: Graph $G = (V, E)$, edge weight $c : E \to \mathbb{R}^+$.

Output: Optimal postman tour ω.

Complexity: **M:** $O(n^{2.5} \log^3 n)$, **L:** $O(n\sqrt{nm} \log^3 n)$ quantum steps.

1: $S := \{v \in V \mid$ number of adjacent vertices of v is odd$\}$

2: **for** $u \in S$ **do**

3:     $P_u :=$ SHORTEST PATH$[G, u, c]$

4: $K :=$ Complete graph on the vertices of S with edge weight $w_{u,v} := |P_{u,v}|$

5: $B :=$ Complete bipartite graph of K

6: $M :=$ MINIMUM WEIGHT PERFECT BIPARTITE MATCHING$[B, w]$

7: $H := G$

8: **for** $\{u, v\} \in M$ **do**

9:     **for** $f \in P_{u,v}$ **do**

10:       $H := H_{+f}$

11: $\omega :=$ Eulerian Tour of H
___

Theorem 7.2.1 *The quantum time complexity of the* OPTIMAL POST-MAN TOUR *algorithm is* $O(n^{2.5} \log^3 n)$ *in the adjacency matrix model and* $O(n\sqrt{nm} \log^3 n)$ *in the adjacency list model.*

Proof. We consider the adjacency matrix model. First we determine the set S of all vertices with odd degree, this can be done in time $O(n^2)$. Then we compute the shortest paths between all the vertices in S. For this part we use the SHORTEST PATH quantum algorithm with running time complexity of $|S| \cdot O(n^{1.5} \log^3 n)$. Let K be the complete graph on all the vertices of S with edge weight $|P_{u,v}|$. The minimum weight perfect matching problem in a complete graph can be transformed into the problem of finding a minimal weight perfect matching of an associate

122

complete bipartite graph. In a bipartite graph, the computation of such a matching can be done in $O(n^{2.5} \log^2 n)$ time steps (see Chapter 6.2). The computation of the graph H and the eulerian tour in H can be computed in $O(n^2)$. In total, the running time complexity for the worst case ($|S| = n$) of the above algorithm is $O(n^{2.5} \log^3 n)$ in the matrix model. The quantum time complexity for the list model follows with similar analysis. □

The algorithm can be applied equally to the digraph version. But for mixed graphs, having both undirected and directed edges the problem becomes **NP**-hart [Pap76].

Theorem 7.2.2 *The optimal postman tour problem requires $\Omega(n^2)$ quantum queries to the adjacency matrix.*

Proof. We define a graph $G = (V, E)$ with n vertices and $m = \Theta(n^2)$ edges. Let $l := \frac{n}{2} - 1$, G satisfies the following requirements:

1. There are two designate vertices s and t which are not connected.

2. There are l mutually not connected red vertices and l mutually not connected blue vertices.

3. The vertex s is connected to every red vertex, and t is connected to every blue vertex. The weight for every edge is zero.

4. The red and blue vertices are connected by either $\frac{l^2}{2}$ or $\frac{l^2}{2} + 1$ edges with weight one chosen at random.

Suppose $\frac{l^2}{2}$ is even. We have to decide whether the weight of the optimal postman tour is $\frac{l^2}{2}$ or $\frac{l^2}{2} + 2$. Therefore we must compute the majority on l^2 bits. There is an $\Omega(l^2)$ quantum query lower bound for majority. Hence the optimal postman tour problem requires $\Omega(n^2)$ quantum queries.

$\square$

From this result follows that with quantum computation we can't speed up the optimal postman tour problem in the query model.

7.3 Hamiltonian Circuit

In this Section, we consider the hamiltonian circuit problem, a well known **NP**-hard problem.

Hamiltonian Circuit (HC): Given a directed graph G, decide if G contains a cycle of length n.

A hamiltonian graph is one containing a hamiltonian cycle. The hamiltonian cycle problem is analogous to the eulerian graph problem, but a simple characterization of a hamiltonian graph does not exist. The hamiltonian circuit problem is a well known **NP**-complete problem.

There is a quantum query lower bound of $\Omega(n^{1.5})$ for the hamiltonian cycle problem in the matrix model, proved by Berzina et al. [BDFLS04]. We show an upper bound for this problem by using the quantum walk search technique.

Theorem 7.3.1 *The quantum query complexity of the hamiltonian cycle* *problem is $O(n^{2n/(n+1)})$ in the adjacency matrix model.*

Proof. We use Theorem 3.3.1. To do so, we construct a Markov chain and a database for checking if a vertex of the chain is marked.

Let $G = (V, E)$ be a directed input graph with n vertices. Let A be a subset of $[n] \times [n]$ of size $r > n$. We will determine r later. The database is the edge-induced subgraph $G[A] := (V, E \cap A)$. Our quantum walk takes place on the Johnson graph $J(n^2, r)$. The marked vertices M of $J(n^2, r)$ correspond to subsets of $[n] \times [n]$ with size r, where $G[A]$ contains a hamiltonian cycle in G for all $A \in M$. In every step of the walk, we exchange one element of A.

We determine the quantum query costs for setup, update and checking. The setup cost for the database is $O(r)$, the update cost is $O(1)$, and the checking cost is zero. The spectral gap of the walk on $J(n^2, r)$ is $\delta = O(1/r)$ for $1 \leq r \leq \frac{n^2}{2}$, see e.g. [BŠ06]. If there is a hamiltonian cycle in G, then there are at least $\binom{n^2-n}{r-n}$ marked sets, since a hamilonian cycle contains n edges. Therefore it holds

$$\varepsilon \geq \frac{|M|}{|X|} \geq \frac{\binom{n^2-n}{r-n}}{\binom{n^2}{r}} \geq \Omega\left(\left(\frac{r}{n^2}\right)^n\right).$$

Then the quantum query complexity of the hamiltonian cycle problem is

$$O\left(r + \left(\frac{n^2}{r}\right)^{n/2} \cdot \sqrt{r}\right) = O(n^{2n/(n+1)})$$

iff $r = n^{2n/(n+1)}$. □

7.4 Travelling Salesman Tour

Here we look for exact and approximation quantum algorithms for the travelling salesman problem.

Travelling Salesman Problem (TSP): Given a weighted graph G, compute a hamiltonian cycle in G with minimum total edge weight.

Bounded Degree Graphs

In this subsection we consider the travelling salesman problem in graphs with maximal degree three, four and five. There is a simple algorithm by Held and Karp [HK62] for computing a travelling salesman tour with running time $O(2^n)$. Today, this is the fastest known algorithm for the TSP.

Eppstein [Epp03] constructed an algorithm for the TSP on graphs with bounded degree three and running time $O(2^{n/3})$. The general idea of this algorithm is the following (see Figure 7.1): Let G be a directed

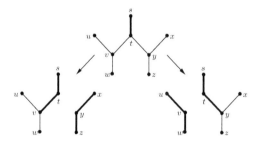

Figure 7.1: Travelling salesman tour in graphs with maximal degree three

weighted graph with maximum degree three. Let F be the set of edges that must be used in the travelling salesman tour, denoted as the *forced edge*. In every step of the algorithm, we choose an edge (t, v) or (t, y) which are adjacent to a forced edge (s, t). If we add (t, v) to F, we delete (t, y) from G, and add the two edges (x, y) and (y, z) to F. Therewith the number of forced edges is increased by three. The subproblem in which we add (t, y) to F is symmetric.

This procedure is the main subroutine of the Eppstein algorithm. It is not difficult to see that we can transform this deterministic algorithm with running time of $O(2^{n/3})$ in a probabilistic polynomial time algorithm with success probability of $1/2^{n/3}$.

From this classical algorithm we obtain a quantum algorithm by the following two modifications: We use Grover search for finding the edges of the graph, and we apply the quantum amplitude amplification [BHMT02] in order to get an algorithm which computes a travelling salesman tour with constant success probability. Then we obtain the following result:

Theorem 7.4.1 *There is a quantum algorithm for the TSP on graphs with bounded degree three and expected running time of $O(2^{n/6})$.*

Now we use an idea of Eppstein [Epp03] to compute the quantum time complexity for finding a travelling salesman tour in graphs with maximal degree four. In classical computation, the fastest algorithm for this problem has running time $O(1.890^n)$, see [Epp03].

Theorem 7.4.2 *There is a quantum algorithm for the TSP on graphs* *with bounded degree four and expected running time of $O((27/4)^{n/6}) = O(1.375^n)$.*

Proof. Let k be the number of degree four vertices in the graph G with maximum degree four. The algorithm consists of the following steps: For each degree four vertex v with adjacent edges a, b, c and d, let a be the incoming edge of the tour. We choose randomly among the three possible partitions $\{a, b\}$, $\{a, c\}$ and $\{a, d\}$. We divide the vertex v into two vertices, and connect the two vertices by a forced edge. The new graph has maximum degree three, and therefore we can apply the quantum algorithm of Theorem 7.4.1.

Each such divide step preserves the travelling salesman tour, if the two edges of the tour do not belong to the same set of the partition. This happens with probability 2/3. We apply the quantum amplitude amplification, and after $\sqrt{(3/2)^k}$ invocations the algorithm finds the correct solution. In total, the quantum time complexity of the algorithm is

$$O\left(1.5^{k/2} \cdot 2^{n/6}\right) = O((27/4)^{n/6}) = O(1.375)^n.$$

□

There is no classical travelling salesman algorithms with running time faster than $O(2^n)$ in graphs with maximal degree five. We present a quantum algorithm for the TSP on graphs with maximal degree five and running time $O(1.5874^n)$. We use the same strategy as for bounded degree four graphs.

Theorem 7.4.3 *There is a quantum algorithm for the TSP on graphs with bounded degree five and expected running time of $O(1.5874^n)$.*

Proof. We use the proof of Theorem 7.4.2. Here we choose randomly among four possible partitions, and divide a vertex with degree five into two vertices, which we connect by a forced edge. Then the new graph has maximum degree four, and we can apply Theorem 7.4.2. In total, the quantum time complexity of the TSP algorithm for bounded degree five is

$$O\left((4/3)^{k/2} \cdot (27/4)^{n/6}\right) = O(1.5874^n).$$

$\square$

By using similar arguments we can construct quantum algorithms for the TSP on graphs with bounded degree six, seven, etc.

Approximation TSP

An application of the results in the previous sections is an approximation quantum query algorithm for TSP. There is a well known classical algorithm for approximation TSP (see for example Gross and Yellen [GY99]), which uses the minimum spanning tree and the eulerian tour problem. This algorithm has an approximation factor 2 that means that the total edge weight of the hamiltonian cycle is never worse than twice the optimal value.

Theorem 7.4.4 *There is an 2-approximation algorithm for TSP with quantum query complexity of $O(n^{1.5})$ in the adjacency matrix model and*

129

$O(\sqrt{nm})$ *in the adjacency list model.*

The proof of the theorem follows by a combination of the classical 2-approximation algorithm for TSP with some quantum graph subroutines (see [Doe07c]).

7.5 Project Scheduling

A digraph model can be used to schedule projects consisting of several interrelated tasks. Some of these tasks can be executed simultaneously, but some tasks cannot begin until certain others are completed. The goal is to compute the minimal project completion time. One way to represent scheduling projects is to use a digraph model, which is called AOA network.

Definition 7.5.1 An *AOA network* $N = (G, c)$ is a digraph $G = (V, E)$ with an edge weight $c : V \times V \rightarrow \mathbb{R}^+$. Each edge in the digraph represents a task of the project, the direction of the edge is the direction of progress in the project. Each vertex in the AOA network represents an event that signifies the completion of one or more activities and the beginning of a new one. An activity A called *predecessor* of activity B, if B cannot begin until A is completed.

We compute the quantum query complexity of the project scheduling problem:

Project Scheduling: Given an AOA network $N = (G, c)$, compute the earliest completion time for every vertex in G.

The AOA network must be an acyclic digraph, otherwise none of the tasks corresponding to the edges on the cycle could ever begin. We are interested on the earliest time point at which each event can occur. Let $ET(i)$ denote the earliest time point in which the event corresponding to vertex i can occur. A vertex j is called *immediate predecessor* of a vertex i if there is an edge from j to i. Let $P(i)$ be the set of all immediate predecessors of vertex i.

Lemma 7.5.2 *It holds* $E(1) = 0$ *and* $E(i) = \max_{j \in P(i)} \{ET(j) + c(j, i)\}$.

The earliest time $ET(i)$ for every event i to occur is the length of the longest directed path in the network from vertex 1 to vertex i.

Theorem 7.5.3 *There is a quantum algorithm for the project scheduling problem with query complexity of* $O(n^{1.5})$ *in the adjacency matrix model and* $O(\sqrt{nm})$ *in the adjacency list model.*

Proof. We use Lemma 7.5.2 to compute the earliest time $ET(i)$ for every vertex $i \in \{2, 3, \ldots, n\}$ in order, since the vertices are numbered in a topological way. We use the quantum algorithm by Dürr and Hoyer [DH96] (see Theorem 3.1.10) to compute the maximum of $ET(j) + c(j, i)$ for all immediate predecessors of vertex i. The quantum query complexity for this step is $O(\sqrt{n})$ in the adjacency matrix model and $O(\sqrt{d_G^-(i)})$ in the adjacency list model. The total number of quantum queries to the adjacency matrix is $O(n^{1.5})$ and

$$\sum_{i=1}^{n} \sqrt{d_{\bar{G}}(i)} \le \sqrt{n} \sqrt{\sum_{i=1}^{n} d_{\bar{G}}(i)} = O(\sqrt{nm})$$

in the adjacency list model. $\qquad\qquad\qquad\qquad\qquad\qquad\qquad$ $\square$

Theorem 7.5.4 *The project scheduling problem requires* $\Omega(n^{1.5})$ *quantum queries to the adjacency matrix and* $\Omega(\sqrt{nm})$ *quantum queries to the adjacency list.*

Proof. The proof is a reduction from maximum finding. Let k be an integer and M be a matrix with n rows, k columns and with $N = kn$ positive entries. The quantum query lower bound for finding the maximum value in every row is $\Omega(\sqrt{nN})$, see [DHHM04].

We construct a weighted graph $G = (V, E)$, where the set of vertices is $V = \{s, v_1, \ldots, v_k, u_1, \ldots, u_n, t\}$. The edges (s, v_i) and (u_j, t) have the weight 0 for all $i \in [k]$ and $j \in [n]$. The edges (v_i, u_j) get the weight M_{ji}. The graph G has $n + k + 2$ vertices and $m = kn + k + n$ edges.

The earliest time $ET(v_i)$ is zero for all vertices v_i, and the earliest time $ET(u_j)$ is the maximal weighted edge of (v, u_i) with $v \in \{v_1, \ldots, v_k\}$. Then the project scheduling problem requires $\Omega(\sqrt{nm})$ quantum queries to the adjacency list. Setting $m = n^2$, the quantum query lower bound for the adjacency matrix follows. $\qquad\qquad$ $\square$

Chapter 8

Independent Set Problems

In this Chapter we present quantum complexity lower and upper bounds for independent set problems. An independent set is a set of vertices of a graph in which no two of these vertices are adjacent. A maximal independent set in a graph is an independent set which is contained in no other independent set. A maximum independent set is a largest independent set of a graph G. The results of this Chapter are published in [Doe07a].

In Section 8.1 we prove quantum query lower and upper bounds for finding a maximal independent set in a graph. We present an $O(\sqrt{nm})$ quantum query algorithm for computing a maximal independent set in a graph. We prove that this quantum algorithm is optimal in the adjacency matrix model. The quantum time complexity of our algorithm is $O(\sqrt{nm}\log^2 n)$, which is better than the best classical algorithm.

In Section 8.2 we present a quantum time algorithms for finding a maximum independent set, an **NP**-hart problem. The development of algorithms for the maximum independent set problems is one of the most

133

applicable problem in graph theory. The maximum independent set problem is closely related to the maximum clique and the minimum vertex cover problem. The first exact algorithms for computing a maximum independent set were given by Tarjan and Trojanowski [TT77] with running time of $O(1.2599^n)$. Jian [Jia86] improved the time complexity to $O(1.2346^n)$, Beigel [Bei99] to $O(1.2227^n)$ and Fomin et al. to $O(1.2209^n)$ [FGK06]. Today, the fastest known algorithm is the one given by Robson [Rob01] with running time of $O(1.1844^n)$. This algorithm is based on a detailed computer generated subcase analysis. We construct an $O(1.1488^n)$ quantum time algorithm for computing a maximum independent set. This algorithm is faster than the best classical algorithm by [Rob01].

8.1 Maximal Independent Set

Definition 8.1.1 A set of vertices $V' \subseteq V$ is called *independent*, if for all distinct vertices $u, v \in V'$ it holds $\{u, v\} \notin E(G)$. The set V' is called *maximal*, if there is no independent set $V'' \subseteq V$ with $V' \subset V''$. A *maximum independent set* is a largest independent set of G. By $\alpha(G)$ we denote the *independence number* of G, i.e. the size of a maximum independent set in G.

We study the quantum query complexity of the following problem:

Maximal Independent Set: Given a graph $G = (V, E)$, compute a maximal independent set in G.

We present an $O(\sqrt{nm})$ quantum query algorithm for computing a maximal independent set in a graph. Then we show that this algorithm is optimal in the adjacency matrix model, by proving a lower bound of $\Omega(n^{1.5})$. Let $G = (V, E)$ be a graph, $S \subseteq V$ and $v \in V$. For the application of quantum search, we define a search function $f_{G,S,v} : V \to \{0, 1\}$ with

$$
f_{G,S,v}(x) := \begin{cases} 1, & \text{if } x \in N_G(v) \text{ and } x \in S \\ 0, & \text{otherwise.} \end{cases}
$$

Algorithm 12 MAXIMAL INDEPENDENT SET

Input: Graph $G = (V, E)$.

Output: Maximal independent set V'.

Complexity: M: $O(n^{1.5} \log n)$, L: $O(\sqrt{nm} \log n)$ quantum queries.

1: $V' := \emptyset$, $S := V$

2: **while** $S \neq \emptyset$ **do**

3:    Choose $v \in S$

4:    $V' := V' \cup \{v\}$

5:    $W :=$ ALL QUANTUM SEARCH$[f_{G,S,v}] \cup \{v\}$

6:    $S := S \backslash W$

Theorem 8.1.2 *The expected quantum query complexity of the* MAXIMAL INDEPENDENT SET *algorithm is* $O(n^{1.5} \log n)$ *in the adjacency matrix model and* $O(\sqrt{nm} \log n)$ *in the adjacency list model.*

Proof.   Let $G = (V, E)$ be a undirected graph, we compute a set V' of independent vertices which is maximal in G. At beginning $V' = \emptyset$ and we mark every vertex of G as enabled. While there are some enabled vertices of G, we choose an enabled vertex v, add it to V' and use Grover search to find all neighbours of v, which we mark as disabled.

In the adjacency matrix model, every vertex is found in $O(\sqrt{n})$ quantum queries to the adjacency matrix. In total we use $O(n^{1.5})$ quantum queries in the adjacency matrix model. In the adjacency list model, processing a vertex v costs $O(\sqrt{d_G(v)a_v})$ quantum queries, where a_v is the number of vertices in F which are adjacent to v. Since $\sum_v a_v \leq n$, then the quantum query complexity is upper-bounded by the Cauchy-Schwarz inequality:

$$\sum_v \sqrt{d_G(v)a_v} \leq \sqrt{\sum_v d_G(v)}\sqrt{\sum_v a_v} = O(\sqrt{mn}).$$

In order to get a constant success probability, we need to amplify the success probability of each subroutine by repeating it $O(\log n)$ times, see Remark 3.1.9. Then we obtain a maximal independent set V' in the indicated quantum query complexity. $\qquad\square$

Corollary 8.1.3 *The expected quantum time complexity of the* MAX-IMAL INDEPENDENT SET *algorithm is* $O(n^{1.5} \log^2 n)$ *in the adjacency matrix model and* $O(\sqrt{nm} \log^2 n)$ *in the adjacency list model.*

Now we prove a $\Omega(n^{1.5})$ quantum query lower bound for the maximal independent set problem with the method of Ambainis [Amb02] and analog to Berzina et al. [BDFLS04]. Consequently the MAXIMAL INDEPENDENT SET quantum algorithm is nearly optimal in adjacency matrix model.

Theorem 8.1.4 *The maximal independent set problem requires* $\Omega(n^{1.5})$
quantum queries to the adjacency matrix.

Proof. We construct the sets A and B for the usage of Theorem 4.1.1.
Let f be the Boolean function which is one, iff there is a maximal independent set of size $2n$. The set A consists of all graphs $G = (V, E)$
with $|V| = 3n + 1$ satisfying the following requirements: 1. There are
n mutually not connected red vertices. 2. There are $2n$ green vertices
not connected with the red ones. Green vertices are grouped in pairs
and each pair is connected by edge. 3. There is a black vertex which is
connected to all red and green vertices. Let V' be the set of n red vertices
and one green vertex of each pairs. Then V' is a maximal independent
set in G. The value of the function f for all graphs $G \in A$ is 1.

The set B consists of all graphs $G' = (V, E)$ with $|V| = 3n + 1$ satisfy-
ing the following requirements: 1. There are $n+2$ mutually not connected
red vertices. 2. There are $2n - 2$ green vertices not connected with red
ones, green vertices are grouped in pairs and each pair is connected by
edge. 3. There is a black vertex which is connected to all red and green
vertices. The value of the function f for all graphs $G' \in B$ is 0, since
there is no maximal independent of size $2n$ in G'.

From each graph $G \in A$, we can obtain $G' \in B$ by deleting one
edge between two green vertices, then $l = n = O(n)$. From each
graph $G' \in B$, we can obtain $G \in A$ by adding an edge between two
red vertices, then $l' = (n + 2)(n + 1)/2 = O(n^2)$. By Theorem 4.1.1, the

quantum query complexity of the maximal independent set problem is
$\Omega(\sqrt{l \cdot l'}) = \Omega(n^{1.5})$. $\square$

8.2 Maximum Independent Set

Now we are interested in the quantum time complexity for computing
a largest independent set in a graph. This is a well known **NP**-hard
problem, which is important for many other applications in computer
science and graph theory.

Maximum Independent Set: Given a graph $G = (V, E)$, compute an
independent set $V' \subseteq V$ with $|V'| = \alpha(G)$.

The first exact algorithms for the maximum independent set problem was given by Tarjan and Trojanowski [TT77] with running time of
$O(1.2599^n)$. Jian [Jia86] improved the time complexity to $O(1.2346^n)$,
Beigel [Bei99] to $O(1.2227^n)$, and Robson [Rob01] to $O(1.1844^n)$. The algorithm by Robson is today the fastest algorithms, it based on a detailed
computer generated subcase analysis (number of subcases is in the tens
of thousands). We construct a quantum algorithm which is faster than
the best classical algorithm. Our quantum algorithm has a running time
of $O(1.1488^n)$. This is no query algorithm, in this algorithm we count
the time steps to compute a maximum independent set. Our algorithm
combines a classical probabilistic algorithm with the quantum amplitude
amplification. First we need two simple facts from the maximal independent set theory.

Lemma 8.2.1 *For a path P_n and a cycle C_n with n vertices, it holds*

$$\alpha(P_n) = \left\lceil \frac{n}{2} \right\rceil \ and \ \alpha(C_n) = \left\lfloor \frac{n}{2} \right\rfloor.$$

Lemma 8.2.2 *Let G be a simple graph with $\Delta(G) \leq 2$. Then all the components of G are paths and cycles.*

With the application of the above two Lemmas we can construct a quantum time algorithm for the maximum independent set problem. If the maximum degree of the graph is at most two, we denote with Paths(G) and Cycles(G) the set of all paths and cycles in graph G. The computation of the maximum independent set of a path or a cycle is then a simple task. We denote with MIS(G') the maximum independent set in a graph G', which is a path or a cycle.

Algorithm 13 MAXIMUM INDEPENDENT SET

Input: A graph $G = (V, E)$.

Output: Maximum independent set (MIS) V'.

Complexity: **M, L:** $O(1.1488^n)$ quantum steps.

1: $F := G$, $V' := \emptyset$

2: **while** $V(F) \neq \emptyset$ **do**

3:    **if** $\Delta(F) \leq 2$ **then**

4:       $V' := \bigcup_{P \in \text{Paths}(F)} \text{MIS}(P) \cup \bigcup_{C \in \text{Cycles}(F)} \text{MIS}(C)$

5:       **return**$[V']$

6:    Find $v \in V(F)$ with $\Delta(F) = \deg_F(v)$

7:    $a \in_R \{0, 1\}$

8:    **if** $a = 0$ **then**

9:       $F := F_{-\{v\}}$

10:   **else**

11:      $V' := V' \cup \{v\}$

12:      $F := F_{-N_F[v]}$

13: Apply AMPLITUDE AMPLIFICATION

Theorem 8.2.3 *The expected quantum time complexity of the* MAXI-
MUM INDEPENDENT SET *algorithm is* $O(2^{n/5}) = O(1.1488^n)$.

Proof. The MAXIMUM INDEPENDENT SET algorithm combines a classical probabilistic algorithm with the quantum amplitude amplification
[BHMT02]. We show that the probability for computing a maximum
independent set with the classical algorithm is at least $\varepsilon = (1/2)^{2n/5}$. To
obtain a quantum algorithm, we just use quantum amplitude amplifica-
tion like [Amb04b]. We search for a largest independent set, which can be
modelled by the maximum finding algorithm by Dürr and Høyer [DH96].

Then we increase the success probability to a constant, by repeating the algorithm $O(\frac{1}{\sqrt{\varepsilon}}) = O(2^{n/5})$ times. Considering this, we obtain the indicated quantum time complexity.

Now we prove that the probability for computing a maximum independent set with the classical algorithm is at least $\varepsilon = (1/2)^{2n/5}$. In the first steps of this algorithm, we check if the maximal degree of the graph is smaller or equal than two. If this is true, we apply Lemma 8.2.1 and Lemma 8.2.2, and compute the maximal independent set V'. Otherwise we choose a vertex v with maximal degree, and a random variable $a \in \{0, 1\}$. If $a = 0$, we assume that v is not in the maximum independent set V', and then we delete the vertex v from F. In the other case, the vertex v is in the maximum independent set V'. We delete v and the set of all neighbours $N_F(v)$ from F. Since $\Delta(G) \geq 3$, we delete at least four vertices.

The task is now to determine the expected number of steps x when F is empty. We choose the value of a with uniform distribution from $\{0, 1\}$. If $a = 0$ we delete one vertex and if $a = 1$ we delete at least four vertices of F, such that $n \geq \frac{1}{2}(1x + 4x)$. Then it is $x \leq 2n/5$ and

$$\text{Prob}(V' \text{ is a MIS}) \geq (1/2)^{2n/5}.$$

Now we apply the amplitude amplification, and repeat the procedure

$$O(1/\sqrt{\text{Prob}(V' \text{ is a MIS})}) = O(2^{n/5}) = O(1.1488^n)$$

times, to compute a maximum independent set V' of G. $\qquad \square$

Theorem 8.2.4 *The maximum independent set problem requires* $\Omega(n^{1.5})$ *quantum queries to the adjacency matrix.*

Proof. Every maximum independent set is a maximal independent set, and this requires $\Omega(n^{1.5})$ quantum queries to the adjacency matrix.

$\square$

The upper bound for the quantum query complexity of the maximum independent set problem follows immediately from graph copy (see Chapter 3.3).

Theorem 8.2.5 *The quantum query complexity of the maximum independent set problem is* $O(n^{2-2/\alpha(G)})$, *where* $\alpha(G)$ *is the size of a maximum independent set in* G.

We can use our maximum independent set algorithm and a decomposition theorem of Raman and Saurabh [RS05] for finding a minimum odd cycle transversal, which is a subset of vertices whose deletion makes the graph bipartite. Therefore we can speed up the classical algorithm by [RS05] from $O(1.62^n)$ to $O(1.58^n)$, see [Doe07a].

Chapter 9

Testing Algebraic Properties

In this Chapter we study the quantum complexity for testing algebraic properties. For a set S and a binary operation on S represented as operation table, we consider the decision problem whether a given structure with the promise of being groupoid, semigroup, monoid or quasigroup is a group. We also present upper and lower bounds for testing associativity, distributivity and commutativity. Testing of algebraic properties are fundamental problems in algebra and computer science, which have many important applications. For example, the verification whether a black box is a group is very useful in cryptography. The results of this Chapter are published in [DT07] [DT08a] and [DT08c].

In Chapter 9.1 we consider the semigroup problem, that is, whether the operation on S is associative. Rajagopalan and Schulman [RS00] developed a randomized algorithm for this problem that runs in time $O(n^2)$. As an additional parameter, we consider the binary operation $\circ : S \times S \to S'$, where $S' \subseteq S$. We construct a quantum algorithm for this problem whose query complexity is $O(n^{5/4})$, if the size of S' is constant.

143

Our algorithm is the first application of the new quantum random walk search scheme by Magniez, Nayak, Roland, and Santha [MNRS07]. With the quantum random walk of Ambainis [Amb04a] and Szegedy [Sze04a], the query complexity of our algorithm would not improve the obvious Grover search algorithm for this problem. Furthermore we show a quantum query lower bound for the semigroup problem of $\Omega(n)$, which holds also if the size of S' is constant.

In Chapter 9.2 we have given a finite set S of size n with a binary operation $\circ : S \times S \to S$ represented by a table. One has to decide whether S has an identity element or is a monoid. We show that the identity problem can be solved with linearly many quantum queries. This is optimal, since we also prove a tight lower bound for this problem. Moreover we show a linear lower bound of the quantum query complexity for testing whether a groupoid is a monoid.

In Chapter 9.3 we consider several group problems. Given a groupoid, semigroup, monoid or quasigroup S by its operation table, we have to decide whether S is a group. We present a randomized algorithm for testing whether a semigroup resp. monoid is a group with running time of $O(n^{\frac{3}{2}})$. This improves the naive $O(n^2)$ algorithm that searches for an inverse in the operation table for every element. Then we show that on a quantum computer the complexity can be improved to $\widetilde{O}(n^{\frac{11}{14}})$. Furthermore we give nearly optimal quantum query algorithms for testing whether a groupoid or quasigroup is a group.

In Chapter 9.4 we present several bounds for testing commutativity. We prove that the quantum query complexity of the commutativity problem for groupoids, semigroups and monoids is $\Theta(n)$. In addition, we show that the commutativity problem can be solved in logarithmic number of quantum queries to the operation table if it is a quasigroup resp. group. In Chapter 9.5 we consider the distributive problem. We have given a set S and two binary operations $\oplus : S \times S \to S$ and $\otimes : S \times S \to S$ represented by a table. One has to decide whether $(S, \oplus, \otimes)$ is distributive, i.e. we have to test whether the two equations $a \otimes (b \oplus c) = (a \otimes b) \oplus (a \otimes c)$ and $(a \oplus b) \otimes c = (a \otimes c) \oplus (b \otimes c)$ are satisfied. We show a linear lower bound on the quantum query complexity for this problem. Moreover we prove that the distributive problem can be decided with linear quantum query complexity if $(S, \oplus)$ is a commutative quasigroup.

9.1 The Semigroup Problem

In the semigroup problem we are given two sets S and $S' \subseteq S$ and a binary operation $\circ : S \times S \to S'$ represented by a table. We denote with n the size of the set S. One has to decide whether S is a semigroup that is, whether the operation on S is associative.

The complexity of this problem was first considered by Rajagopalan and Schulman [RS00], who gave a randomized algorithm with time complexity of $O(n^2 \log \frac{1}{\delta})$, where δ is the error probability. They also showed a lower bound of $\Omega(n^2)$. The previously best known algorithm was the naive $O(n^3)$-algorithm that checks all triples.

In the quantum setting, one can do a Grover search over all triples $(a, b, c) \in S^3$ and check whether the triple is associative. The quantum query complexity of the search is $O(n^{3/2})$. We construct a quantum algorithm for the semigroup problem that has query complexity $O(n^{5/4})$, if the size of S' is constant. Furthermore we give a quantum query lower bound of $\Omega(n)$ for this problem. Our algorithm is the first application of the recent quantum random walk search scheme by Magniez et al. [MNRS07]. The quantum random walk of Ambainis [Amb04a] and Szegedy [Sze04a] doesn't suffice to get an improvement of the Grover search mentioned above.

Theorem 9.1.1 *Let* $k = n^\alpha$ *be the size of* S' *with* $0 < \alpha \le 1$. *The quantum query complexity of the semigroup problem is*

$$
\begin{cases}
O(n^{\frac{5+\alpha}{4}}), & \text{for } 0 < \alpha \le \frac{1}{3}, \\
O(n^{\frac{6+2\alpha}{5}}), & \text{for } \frac{1}{3} < \alpha \le \frac{3}{4}, \\
O(n^{\frac{3}{2}}), & \text{for } \frac{3}{4} < \alpha \le 1.
\end{cases}
$$

Proof. We use the quantum walk search scheme of Theorem 3.3.1. To do so, we construct a Markov chain and a database for checking if a vertex of the chain is marked.

Our quantum walk is done on the categorical graph product G_J of two Johnson graphs $J(n, r)$. Let A and B two subsets of S of size r. We will determine r later. We search for a pair $(a, b) \in S^2$, such that a, b are two elements of a nonassociative triple. Then the marked vertices of G_J correspond to pairs (A, B) with $(A \circ B) \circ S \ne A \circ (B \circ S)$. In every step of the walk, we exchange one row and one column of A and B.

The database of our quantum walk is the set

$$D(A, B) = \{ (a, b, a \circ b) \mid a \in A \cup S' \text{ and } b \in B \cup S' \}.$$

Now we compute the quantum query costs for the setup, update and checking. The setup cost for the database $D(A, B)$ is $O((r + k)^2)$ and the update cost is $O(r + k)$.

To check whether a pair (A, B) is marked, we have to test if $(A \circ B) \circ S \neq A \circ (B \circ S)$. We claim that the quantum query cost to check this inequality is $O(\sqrt{nrk})$: fix a pair $(b, c) \in B \times S$. We check whether $(A \circ b) \circ c \neq A \circ (b \circ c)$. To get $(b \circ c)$ we make one query to the oracle. Because $(b \circ c) \in S'$, the computation of $A \circ (b \circ c)$ can be done by using our database. We obtain r values which we denote by $(y_1, \ldots, y_r)$. The evaluation of $(A \circ b)$ needs no queries by using our database, let $(z_1, \ldots, z_r)$ be the result. Note that $z_i \in S'$ for all i. Now we use Grover's algorithm for searching a $s \in S'$ such that $s \circ c \neq y_j$ for a $j \in [r]$ with $s = z_j$. This search can be done in $O(\sqrt{k})$ quantum queries. The outer loop is a Grover search for a pair $(b, c) \in B \times S$. Therefore, the total checking cost is $O(\sqrt{nrk})$.

The spectral gap of the walk on G_J is $\delta = O(1/r)$ for $1 \leq r \leq \frac{n}{2}$, see [BŠ06]. If there is a triple (a, b, c) with $(a \circ b) \circ c \neq a \circ (b \circ c)$, then there are at least $\binom{n-1}{r-1}^2$ marked sets (A, B). Therefore we have

$$\varepsilon \geq \frac{|M|}{|X|} \geq \left(\frac{\binom{n-1}{r-1}}{\binom{n}{r}} \right)^2 = \frac{r^2}{n^2}.$$

Let $r = n^\beta$ for $0 < \beta < 1$. Assuming $r > k$, then the quantum query complexity of the semigroup problem is

$$O\left(r^2 + \frac{n}{r}\left(\sqrt{r} \cdot r + \sqrt{nrk}\right)\right) = O\left(n^{2\beta} + n^{1+\frac{\beta}{2}} + n^{\frac{3+\alpha-\beta}{2}}\right).$$

Now we choose β depending on α such that this expression is minimal. Suppose that $2\beta \le 1 + \frac{\beta}{2}$, i.e. $\beta \le \frac{2}{3}$. From the equation $1 + \frac{\beta}{2} = \frac{3+\alpha-\beta}{2}$, we get $\beta = \frac{1+\alpha}{2}$. Then the quantum query complexity of the semigroup problem is $O(n^{\frac{5+\alpha}{4}})$ for $r = n^{\frac{1+\alpha}{2}}$ and $\alpha \le \frac{1}{3}$. Otherwise if $2\beta > 1 + \frac{\beta}{2}$, i.e. $\beta > \frac{2}{3}$, we get $\beta = \frac{3+\alpha}{5}$ from the equation $2\beta = \frac{3+\alpha-\beta}{2}$. Then the quantum query complexity is $O(n^{\frac{6+2\alpha}{5}})$ for $r = n^{\frac{3+\alpha}{5}}$ and $\alpha > \frac{1}{3}$. If $\alpha > \frac{3}{4}$, the query complexity is bigger than $O(n^{\frac{3}{2}})$, therefore we use Grover search instead of quantum walk search. $\qquad\square$

For the special case that $\alpha = 0$, i.e., only a constant number of elements occurs in the operation table, we get

Corollary 9.1.2 *The quantum query complexity of the semigroup problem is $O(n^{\frac{5}{4}})$, if S' has constant size.*

Note that the time complexity of our algorithm is $O(n^{1.5} \log n)$.

Theorem 9.1.3 *The semigroup problem requires $\Omega(n)$ quantum queries.*

Proof. Let S be a set of size n and $\circ : S \times S \to \{0,1\}$ a binary operation represented by a table. We apply Theorem 4.1.1. The set A consists of all $n \times n$ matrices, where the entry of position $(1,1)$, $(1,c)$, $(c,1)$ and (c,c) is 1, for $c \in S - \{0,1\}$, and zero otherwise. It is easy to see, that

148

the operation tables of A are associative, since $(x \circ y) \circ z = x \circ (y \circ z) = 1$ for all $x, y, z \in \{1, c\}$ and zero otherwise.

The set B consists of all $n \times n$ matrices, where the entry of position $(1, 1)$, $(1, c)$, $(c, 1)$, (c, c) and (a, b) is 1, for fixed $a, b, c \in S - \{0, 1\}$ with $a, b \neq c$, and zero otherwise. Then $(a \circ b) \circ c = 1$ and $a \circ (b \circ c) = 0$. Therefore the operation tables of B are not associative.

From each $T \in A$, we can obtain $T' \in B$ by replacing the entry 0 of T at (a, b) by 1, for any $a, b \notin \{0, 1, c\}$. Hence we have $m = \Omega(n^2)$. From each $T' \in B$, we can obtain $T \in A$ by replacing the entry 1 of T' at position (a, b) by 0, for $a, b \notin \{0, 1, c\}$. Then we have $m' = 1$. By Theorem 4.1.1, the quantum query complexity is $\Omega(\sqrt{m \cdot m'}) = \Omega(n)$.

$\square$

From our proof follows that the lower bound holds also for constant size of S'.

9.2 The Monoid Problem

In the monoid problem we are given a finite set S of size n with a binary operation $\circ : S \times S \rightarrow S$ represented by a table. One has to decide whether S is a monoid.

The monoid problem is an extention of the semigroup problem of the previous section. We have to verify whether the groupoid $(S, \circ)$ is associative and has an identity element. At first we consider the identity problem, i.e. we have to decide whether there is an identity element. We show that the identity problem requires linearly many quantum queries. We start by considering the 1-column problem: given a 0-1-matrix of order n, decide whether it contains a column that is all 1.

Lemma 9.2.1 *The 1-column problem requires $\Omega(n)$ quantum queries.*

Proof. We use Theorem 4.1.1. The set A consists of all matrices, where in $n - 1$ columns there is exactly one entry with value 0, and the other entries of the matrix are 1. The set B consists of all matrices, where in every column there is exactly one entry with value 0, and the other entries of the matrix are 1. From each matrix $T \in A$, we can obtain $T' \in B$ by changing one entry in the 1-column from 1 to 0. Then we have $m = n$. From each matrix $T' \in B$, we can obtain $T \in A$ by changing one entry from 0 to 1. Then we have $m' = n$. By Theorem 4.1.1, the quantum query complexity is $\Omega(n)$. $\square$

Theorem 9.2.2 *The identity problem requires $\Omega(n)$ quantum queries.*

Proof. We reduce the 1-column problem to the identity problem. Given a 0-1-matrix $M = (m_{i,j})$ of order n. We define $S = \{0, 1, \ldots, n\}$ and a operation table $T = (t_{i,j})$ with $0 \le i, j \le n$ for S as follows:

$$
t_{i,j} = \begin{cases} 0, & \text{if } m_{i,j} = 0, \\ i, & \text{if } m_{i,j} = 1, \end{cases}
$$

and $t_{0,j} = t_{i,0} = 0$. Then M has a 1-column iff T has an identity element. $\square$

Finding an identity element is simple. We choose an element $a \in S$ and then we test if a is the identity element by using Grover search in $O(\sqrt{n})$ quantum queries. The success probability of this procedure is $\frac{1}{n}$. By using the amplitude amplification we get an $O(n)$ quantum query algorithm for finding an identity element (if there is one). Since the upper and the lower bound match, we have determined the precise complexity of the identity problem.

Corollary 9.2.3 *The quantum query complexity of the identity problem is $\Theta(n)$.*

Theorem 9.2.4 *Whether a groupoid is a monoid requires $\Omega(n)$ quantum queries.*

Proof. We reduce the semigroup problem to the monoid problem for groupoids. Let $(S, \circ)$ be a groupoid represented as a operation table. We define a groupoid $M = S \cup \{e\}$ with the identity element $e \notin S$, that is, with $a \circ e = e \circ a = a$, for all $a \in S$. Then the groupoid $(S, \circ)$ is a semigroup iff $(M, \circ)$ is a monoid. □

9.3 The Group Problem

In this section we consider the decision problems whether a given structure with the promise of being a groupoid, semigroup, monoid or quasigroup S of size n with a binary operation $\circ$ is in fact a group.

Group testing for Monoids

We consider the problem whether a given finite monoid M is in fact a group. That is, we have to check whether every element of M has an inverse. The monoid M has n elements and is given by its operation table and the identity element e.

To the best of our knowledge, this special group problem has not been studied before. The naive approach for the problem checks for every element $a \in M$, whether e occurs in a's row in the operation table. The query complexity is $O(n^2)$. We develop a (classical) randomized algorithm that solves the problem with $O(n^{\frac{3}{2}})$ queries to the operation table. Then we show that on a quantum computer the query complexity can be improved to $\widetilde{O}(n^{\frac{11}{14}})$.

Theorem 9.3.1 *Whether a given monoid is a group can be decided with*

1. *$O(n^{\frac{3}{2}})$ queries by a randomized algorithm.*

2. *$O(n^{\frac{11}{14}} \log n)$ by a quantum query algorithm.*

At first we present our classical algorithm, then we prove the correctness of this algorithm, and at last we show how to speed up this algorithm by quantum tools.

Algorithm 14 GROUP TESTING

Input: Multiplication table of a monoid $(M, \circ)$.

Output: 1, if $(M, \circ)$ is a group; 0 otherwise.

1: $i := 0$, $r := n^{1/2}$

2: **for** $a \in M$ **do**

3:     Compute $S_r(a) := (a, a^2, \ldots, a^r)$

4:     **if** $e \in S_r(a)$ **then**

5:         $i := i + 1$

6:     **else if** $\exists x, y \in S_r(a) : x = y$ **then**

7:         **return**[0]

8: **if** $i = n$ **then**

9:     **return**[1]

10: **while** n/r **do**

11:     Choose $a \in_R M$

12:     **if** a has no inverse **then**

13:         **return**[0]

14: **return**[1]

Proposition 9.3.2 *The* GROUP TESTING *algorithm accepts with probability 1 if the monoid $(M, \circ)$ is a group, otherwise it rejects with constant probability.*

Proof.   Let $a \in M$, we consider the sequence of powers $a, a^2, a^3, \ldots$. Since M is finite, there will be a repetition at some point. We define the *order of* a as the smallest power t, such that $a^t = a^s$, for some $s < t$. Clearly, if a has an inverse, s must be zero.

Lemma 9.3.3 *Let $a \in M$ of order t. Then a has an inverse iff $a^t = e$.*

Hence the powers of a will tell us at some point whether a has an inverse.

On the other hand, if a has no inverse, the powers of a provide more elements with no inverse as well.

Lemma 9.3.4 *Let $a \in M$. If a has no inverse, then a^k has no inverse, for all $k \geq 1$.*

Our algorithm has two phases. In phase 1, it computes the powers of every element up to certain number r. That is, we consider the sequences $S_r(a) = (a, a^2, \ldots, a^r)$, for all $a \in M$. If $e \in S_r(a)$ then a has an inverse by Lemma 9.3.3. Otherwise, if we find a repetition in the sequence $S_r(a)$, then, again by Lemma 9.3.3, a has no inverse and we are done.

If we are not already done by phase 1, i.e. there are some sequences $S_r(a)$ left such that $e \notin S_r(a)$ and $S_r(a)$ has pairwise different elements, then the algorithm proceeds to phase 2. It selects some $a \in M$ uniformly at random and checks whether a has an inverse by searching for e in the row of a in the operation table. This step is repeated n/r times.

For the correctness observe that the algorithm accepts with probability 1 if M is a group. Now assume that M is not a group. Assume further that the algorithm does not already detect this in phase 1. Let a be some element without an inverse. By Lemma 9.3.3, the sequence $S_r(a)$ has r pairwise different elements which don't have inverses too by Lemma 9.3.4. Therefore in phase 2, the algorithm picks an element without an inverse with probability of at least r/n. By standard arguments, the probability that at least one out of n/r many randomly chosen elements has no inverse, is constant. $\qquad\square$

Proof of Theorem 9.3.1. First we determine the classical query complexity. The query complexity of the algorithm is bounded by rn in phase 1 and by n^2/r in phase 2. Total the query complexity of the algorithm is

$$O\left(nr + n^2/r\right),$$

which is minimized for $r = n^{\frac{1}{2}}$. Hence the query complexity for testing if a semigroup is a group, is $O(n^{\frac{3}{2}})$.

For the quantum query complexity we use Grover search and amplitude amplification. In phase 1, we search for an $a \in M$, such that the sequence $S_r(a)$ has r pairwise different entries different from e. This property can be checked by first searching $S_r(a)$ for an occurance of e by a Grover search with $\sqrt{r} \log r$ queries. Then, if e doesn't occur in $S_r(a)$, we check whether there is an element in $S_r(a)$ that occurs more than once. This is the element distinctness problem and can be solved with $r^{2/3} \log r$ queries, see [Amb04a]. Therefore the quantum query complexity of phase 1 is bounded by $\sqrt{n} \cdot r^{2/3} \log r$. In phase 2 we search for an $a \in M$ such that a has no inverse. Therefore we actually search the row of a in the operation table. Hence this takes $\sqrt{n}$ queries. Since at least r of the a's don't have an inverse, by amplitude amplification we get $\sqrt{n}\sqrt{n/r} = n/\sqrt{r}$ queries in phase 2. In summary, the quantum query complexity is

$$O(\sqrt{n} \cdot r^{2/3} \log r + \frac{n}{\sqrt{r}}),$$

which is minimized for $r = n^{\frac{3}{7}}$. Hence we have a $O(n^{\frac{11}{14}} \log n)$ quantum query algorithm. □

The time complexity of our classical algorithm is $O(n^{\frac{3}{2}})$. Our quantum implementation have nearly quadratic speed up over the classical algorithm. From the analysis of the algorithm follows that we have used several Grover search subroutines, one amplitude amplification and one application of the quantum walk element distinctness procedure by Ambainis [Amb04a]. Therefore the quantum time complexity is $O(n^{\frac{11}{14}} \log^c n)$ for a constant c, since the element distinctness procedure has running time of $O(n^{\frac{2}{3}} \log^c n)$.

Corollary 9.3.5 *The running time complexity of the* GROUP TESTING *algorithm is $O(n^{\frac{3}{2}})$ in classical setting and $O(n^{\frac{11}{14}} \log^c n)$ in quantum setting.*

Group testing for Semigroups

Now we consider the problem whether a finite semigroup $(S, \circ)$ is in fact a group. The naive approach for this problem searches first for an identity element e of S and then checks whether e occurs in every row of the operation table. The query complexity of this procedure is $O(n^2)$, resp. $O(n)$ in the quantum case.

Theorem 9.3.6 *Whether a given semigroup is a group can be decided with*

1. *$O(n^{\frac{3}{2}})$ queries by a randomized algorithm.*

2. *$O(n^{\frac{11}{14}} \log n)$ by a quantum query algorithm.*

Proof. Our input is a finite semigroup $(S, \circ)$, and we want to decide whether it is in fact a group. To do so, we first search for an identity element and then use the algorithm of Theorem 9.3.1. To find the identity element, we start by choosing an element a of S and search for an element $e \in S$ such that $a \circ e = a$. Then e is our candidate for the identity element. Recall that we finally want to decide whether S is a group. In this case, the identity element is unique. Hence if our candidate e doesn't work we can safely reject the input, even in the case that S actually has an identity element. To test our candidate e, it suffices to check whether $b \circ e = b$ for all $b \in S$. Obviously the two steps can be done in $O(n)$ queries classically and $O(\sqrt{n})$ quantum queries with Grover search.

$\square$

The result should be contrasted with the following: if we want to decide whether a given semigroup is in fact a monoid, then the best known algorithms make $O(n^2)$ queries classically and $O(n)$ queries in the quantum setting.

Group testing for Quasigroups

Next we assume that the input $(S, \circ)$ is a quasigroup. Rajagopalan and Schulman [RS00] showed that in a quasigroup we can deterministically compute a set of generators of size $\log n$ in quadratic time. Light observed (see [CP61]) that if $R \subset S$ is a set of generators of S, then it suffices to test all triples a, b, c in which b is an element of R. Therefore Light's observation results in an $O(n^2 \log n)$ deterministic algorithm for verifying associativity of quasigroups.

Theorem 9.3.7 *Whether a given quasigroup or a loop is a group can be decided with expected quantum query complexity of $\Theta(n)$.*

Proof. First we prove the upper bound. We have to verify if the quasigroup $(S, \circ)$ is associative. Therefore we choose three elements $a, b, c \in S$, and then we verify if $(a \circ b) \circ c \neq a \circ (b \circ c)$. Rajagopalan and Schulman [RS00] showed that any nonassociative quasigroup has at least $n - 2$ nonassociative triples. Then the success probability for finding a nonassociative triple (if there is one) is at least $\frac{n-2}{n^3}$. By using the quantum amplitude amplification we have an $O(n)$ quantum query algorithm for finding a nonassociative triple in a quasigroup (if there is one).

For the lower bound, we apply Theorem 4.1.3 in connection with an idea of [RS00] for proving an $\Omega(n^2)$ lower bound for this problem in classical computing. The set A consists of the operation table T of the group $(\mathbb{Z}_2^m, +)$, where $+$ is the vector addition modulo 2. Let $a, b, c \in \mathbb{Z}_2^m$ with $a \neq 0$. The set B consists of all operation tables of $(\mathbb{Z}_2^m, \circ)$, where $\circ$ is equal to $+$ except in the following four positions:

1. $b \circ c = b + (a + c)$,

2. $b \circ (a + c) = b + c$,

3. $(a + b) \circ c = b + c$,

4. $(a + b) \circ (a + c) = a + b + c$.

158

All tables of B are quasigroups, because the above modifications simply exchange two elements in two rows of the table T, but they are not associative, since

$$a + b = (c \circ (a + b)) \circ c \neq c \circ ((a + b) \circ c) = b.$$

The relation R is defined by

$R = \{\, (T, T') \in (A, B) \mid T' \text{ originates of the above four modifications of } T \,\}.$

Then R satisfies $m = \Omega(n^3)$, $m' = 1$, $l = \Omega(n)$ and $l' = 1$. $\qquad\square$

Group testing for Groupoids

Now we consider the problem whether an arbitrary $(S, \circ)$ is in fact a group. There is a $O(n^2 \log n)$ deterministic algorithm for this problem by [RS00]. We develop a quantum algorithm that has time complexity $O(n^{\frac{13}{12}} \log^2 n)$. Furthermore, we present an $O(n \log n)$ query algorithm for this problem, that has a time complexity $O(n^2 \log n)$ however. The latter algorithm is nearly optimal with respect to the query complexity, as we prove a linear lower bound for this problem.

We need a generalization of a lemma from [RS00]. First we generalize the notion of a cancellative operation.

Definition 9.3.8 Let $(S, \circ)$ be a groupoid with n elements represented by its operation table T. Let $I, J \subseteq [n]$ be two index sets and let $T_{I,J}$ be the subtable of T indexed by I and J. We call $\circ$ *cancellative on* $T_{I,J}$, if every element occurs at most once in every row and every column of $T_{I,J}$.

Lemma 9.3.9 [RS00] *Let* $\circ$ *be cancellative on a* $r \times n$ *subtable of its operation table. If* $\circ$ *is non-associative, then it has at least* $r/4$ *non-associative triples.*

Proof. Let (a, b, c) be a non-associative triple and $a = a' \circ a''$. Consider the following cycle of equations:

$$(a' \circ a'') \circ (b \circ c) = ((a' \circ a'') \circ b) \circ c$$
$$= (a' \circ (a'' \circ b)) \circ c$$
$$= a' \circ ((a'' \circ b)) \circ c))$$
$$= a' \circ (a'' \circ (b \circ c))$$
$$= (a' \circ a'') \circ (b \circ c).$$

Every equation is an application of the associativity law. Since (a, b, c) is a non-associative triple, the first equation fails. Therefore at least one of the other equations must fail as well. Hence at least one of the following four triples must be non-associative:

1. (a', a'', b),

2. $(a', a'' \circ b, c)$,

3. (a'', b, c),

4. $(a', a'', b \circ c)$.

If $\circ$ is cancellative on a $r \times n$ subtable, then a can be written as $a' \circ a''$ in r different ways. Then the associativity fails in at least one of the four categories for each of these r pairs. Hence there is a category for which there are at least $r/4$ failures. Since each category identifies either a' or a'', there are no duplicate triples in any category. $\qquad\qquad\square$

Theorem 9.3.10 *Whether a groupoid is a group can be decided by a quantum algorithm with $O(n^{\frac{13}{12}} \log^c n)$ expected steps, for some constant c.*

Proof. Let $(S, \circ)$ be a groupoid represented by its operation table T. Recall that if S is a group, then $\circ$ is cancellative. Our first step is to determine whether the operation is associative. To do so, we choose an arbitrary subset A of S of size r. We determine r later. Then we check whether $\circ$ is cancellative on the subtable T_A of T, where T_A is the $r \times n$ table that consists of the rows of T indexed by A. This is not the case, if we find a row or column in T_A with two equal elements. Hence we can solve this with a Grover search and the element distinctness quantum algorithm by Ambainis [Amb04a]. The quantum query complexity of this procedure is $O(\sqrt{r}n^{\frac{2}{3}} + \sqrt{n}r^{\frac{2}{3}})$.

If any of the considered rows and columns are not cancellative then we are done. Otherwise we randomly choose three elements $a, b, c \in S$ and check whether $(a \circ b) \circ c \neq a \circ (b \circ c)$. If the operation is not associative, then the probability of finding a non-associative triple is at least $\frac{r}{4n^3}$ by Lemma 9.3.9. By using the quantum amplitude amplification we have an $O(n^{\frac{3}{2}}/\sqrt{r})$ quantum query algorithm for finding a non-associative triple.

If there is no non-associative triple, then $(S, \circ)$ is a semigroup. Whether this semigroup is a group can be decided with $O(n^{\frac{11}{14}} \log n)$

quantum queries by Theorem 9.3.6. The total expected quantum query complexity of this algorithm we get

$$O\left(\sqrt{r}n^{\frac{2}{3}} + \sqrt{n}r^{\frac{2}{3}} + \frac{n^{\frac{3}{2}}}{\sqrt{r}} + n^{\frac{11}{14}}\log n\right).$$

This expression is minimized for $r = n^{\frac{5}{6}}$. Hence the expected time complexity of this algorithm is $O(n^{\frac{13}{12}}\log^c n)$ for a constant c. $\qquad\square$

By setting $r = n$ in Lemma 9.3.9, we have

Corollary 9.3.11 *Whether a groupoid is a quasigroup can be decided by a quantum algorithm within $O(n^{\frac{7}{6}}\log n)$ expected steps.*

We can further improve the query complexity of the problem, if we allow a larger running time.

Theorem 9.3.12 *Whether a groupoid is a group can be decided with $O(n\log n)$ expected quantum queries.*

Proof. Let $(S, \circ)$ be a groupoid represented by its operation table T. A well known fact from algebra is, that if $(S, \circ)$ is a quasigroup, then a random subset $R \subset S$ with $c\log n$ elements is a set of generators with probability at least $1 - \exp(c)$ (see [RS00]). We choose a random subset R of $O(\log n)$ elements of S. Then we check whether R is a generating set of $(S, \circ)$. To do so, let $S_0 = R$. We compute inductively $S_i = S_{i-1} \cup (R \circ S_{i-1})$. This adds at least one element in a step, until we reach some $k \le n$ such that $S_k = S$. In this case, R is a set of generators. For each element a added to some set S_i, we query the $\log n$

elements $R \circ a$ to look for further elements. In total we query at most the $O(n \log n)$ elements of the $R \times S$ submatrix of T. The quantum time is bounded by $O(n^{3/2} \log n)$.

If R is a set of generators, we have to verify whether the multiplication table is associative. Light observed (see [CP61]) that if R is a set of generators of S, then it suffices to test all triples a, b, c in which b is an element of R. By using Grover search, the quantum query for finding a nonassociative triple (if there is one) is $O(n\sqrt{\log n})$. By Theorem 9.3.6 we can decide whether this semigroup is a group. The total quantum query complexity of is $O(n \log n)$. □

The upper bound of Theorem 9.3.12 almost matches the lower bound we have.

Theorem 9.3.13 *Whether a groupoid is a quasigroup or a group requires* $\Omega(n)$ *quantum queries.*

Proof. We apply the Theorem 4.1.3. Let A be the operation table T of $\mathbb{Z}_n$ and let $\circ$ be the addition modular n. Then T is a quasigroup resp. group. The set B consists of all $n \times n$ matrices T', where one entry of T' is modified. Therefore the tables of B forming no quasigroup resp. group. The relation R is defined by

$$R = \{ (T, T') \in (A, B) \mid d(T, T') = 1 \}.$$

Then R satisfies that $m = n^2(n-1)$, $m' = 1$, $l = n-1$ and $l' = 1$. Therefore the quantum query complexity to decide whether a groupoid is a quasigroup resp. group is $\Omega(n)$. $\qquad\qquad\qquad\qquad\square$

9.4 The Commutativity Problem

In the commutativity problem we are given a finite set S of size n with a binary operation $\circ : S \times S \to S$ represented by a table. One has to decide whether S is a commutative. In the quantum setting, one can solve the problem in linear time by a Grover search over all tuples $(a, b) \in S^2$ that checks whether the tuples are commutative. We show that the commutativity problem requires $\Omega(n)$ quantum queries, even when S is a monoid.

Theorem 9.4.1 *The quantum query complexity of the commutativity problem for groupoids, semigroups, and monoids is $\Theta(n)$.*

Proof. We start by showing the lower bound for semigroups via Theorem 4.1.3. Let $S = \{0, 1, \dots, n - 1\}$. The set A consists of the zero matrix of order n. The set B consists of all $n \times n$ matrices, where the entry of position (a, b) is 1, for $a \neq b \in S - \{0, 1\}$, and 0 otherwise. All operation tables of the sets A and B are semigroups. Then we have $m = \Omega(n^2)$, $m' = 1$, and the quantum query lower bound for testing if a given semigroup is commutative is $\Omega(n)$.

We reduce the commutativity problem for semigroups to the commutativity problem for monoids. Let S be a semigroup represented as a operation table T. We define a monoid $M = S \cup \{e\}$ with the identity element $e \notin S$, that is, with $a \circ e = e \circ a = a$, for all $a \in S$. Then the semigroup S is commutative iff the monoid M is commutative. $\qquad\square$

Magniez and Nayak [MN05] quantize a classical Markov chain for testing the commutativity of a black box group given by the generators. They constructed an $O(k^{2/3} \log k)$ quantum query algorithm, where k is the number of generators of the group. In the case when $(S, \circ)$ is a quasigroup, a random set of $c \log n$ elements will be a set of generators with probability at least $1 - \exp(c)$ [RS00]. Therefore we obtain the following result:

Theorem 9.4.2 *Whether a quasigroup, loop or group is commutative can be decided with quantum query complexity* $O((\log n)^{\frac{2}{3}} \log \log n)$.

9.5 The Distributivity Problem

In the distributivity problem we are given a set S and two binary operations $\oplus : S \times S \to S$ and $\otimes : S \times S \to S$ represented by tables. One has to decide whether $(S, \oplus, \otimes)$ is distributive, i.e. we have to test whether the two equations $a \otimes (b \oplus c) = (a \otimes b) \oplus (a \otimes c)$ and $(a \oplus b) \otimes c = (a \otimes c) \oplus (b \otimes c)$ are satisfied. A triple $(a, b, c) \in S^3$ that fulfills both equations is called a *distributive triple*. In classical computing, it is not known whether this problem can be solved in less than cubic time. In the quantum setting, one can do a Grover search over all triples $(a, b, c) \in S^3$ and check whether each triple is distributive. The quantum query complexity of the search is $O(n^{3/2})$. We show a linear lower bound on the query complexity.

Theorem 9.5.1 *The distributivity problem requires $\Omega(n)$ quantum queries.*

Proof. Let $S = \{0, 1, \ldots, n-1\}$. We apply the Theorem 4.1.1. The set A consists of all pairs of $n \times n$ matrices $T_\oplus$ and $T_\otimes$, where $T_\otimes$ is the zero-matrix, and the entry at position $(1,0)$ in $T_\oplus$ is 1, and 0 otherwise. It is easy to see that the tables of A are distributive, since $x \otimes (y \oplus z) = (x \otimes y) \oplus (x \otimes z) = 0$ for all $x, y, z \in S$. The set B consists of all pairs of $n \times n$ matrices $T'_\oplus$ and $T'_\otimes$, where the entry of position $(1,0)$ in $T'_\oplus$, and (a, b) in $T'_\otimes$ is 1, for $a, b \in S - \{0, 1\}$, and 0 otherwise. Then $a \otimes (b \oplus c) = 0$ and $(a \otimes b) \oplus (a \otimes c) = 1$ with $b \neq c$. Therefore the tables of B are not distributive.

From each $(T_\oplus, T_\otimes) \in A$, we can obtain $(T'_\oplus, T'_\otimes) \in B$ by replacing the entry 0 of $T_\otimes$ at (a, b) by 1, for any $a, b \notin \{0, 1\}$. Hence we have $m = \Omega(n^2)$. From each $(T'_\oplus, T'_\otimes) \in B$, we can obtain $(T_\oplus, T_\otimes) \in A$, by replacing the entry 1 of $T_\otimes$ at position (a, b) by 0, for $a, b \notin \{0, 1\}$. Thus we have $m' = 1$. By Theorem 4.1.1, the quantum query complexity is $\Omega(\sqrt{m \cdot m'}) = \Omega(n)$. $\qquad\square$

If $(S, \oplus)$ is a commutative quasigroup, then we can get a faster algorithm to check distributivity. The key is that one nondistributive triple implies the existence of more such triples. Similar to Lemma 9.3.9, we have the following lemma.

Lemma 9.5.2 *Let S be a set and $\oplus, \otimes$ be two binary operations on S, such that $(S, \oplus)$ is a commutative quasigroup. If $(S, \oplus, \otimes)$ is nondis-*

tributive, then it has at least $\Omega(n)$ nondistributive triples.

Proof. Let (a, b, c) be a nondistributive triple. Let $a = a' \oplus a''$ and consider the following cycle.

$$
\begin{aligned}
(a' \oplus a'') \otimes (b \oplus c) &= ((a' \oplus a'') \otimes b) \oplus ((a' \oplus a'') \otimes c) \\
&= ((a' \otimes b) \oplus (a'' \otimes b)) \oplus ((a' \oplus a'') \otimes c) \\
&= (a' \otimes b) \oplus (a'' \otimes b) \oplus (a' \otimes c) \oplus (a'' \otimes c) \\
&= (a' \otimes b) \oplus (a' \otimes c) \oplus (a'' \otimes (b \oplus c)) \\
&= (a' \otimes (b \oplus c)) \oplus (a'' \otimes (b \oplus c)) \\
&= (a' \oplus a'') \otimes (b \oplus c).
\end{aligned}
$$

Suppose that $a \otimes (b \oplus c) \neq (a \otimes b) \oplus (a \otimes c)$. Then at least one of the above equations does not hold. Therefore at least one of the following triples must be nondistributive:

1. (a', a'', b),   3. (a'', b, c),   5. $(a', a'', b \oplus c)$.

2. (a', a'', c),   4. (a', b, c),

Since $(S, \oplus)$ is a quasigroup, a can be written as $a' \oplus a''$ in n different ways. For each of these, distributivity fails in at least one of the five categories from above. Therefore there exists a category for which there are $\geq n/5$ failures.

The case that $(a \oplus b) \otimes c \neq (a \otimes c) \oplus (b \otimes c)$ can be handled similarly

□

By using Lemma 9.5.2 in combination with the amplitude amplification (similar to Theorem 9.3.1) we have

Theorem 9.5.3 *Let $(S, \oplus)$ be a commutative quasigroup and $(S, \otimes)$ a groupoid. Whether $(S, \oplus, \otimes)$ is distributive can be decided with expected quantum query complexity of $O(n)$.*

Chapter 10

Linear Algebra Problems

In this Chapter we study the quantum complexity of some important linear algebra problems. We consider matrix multiplication, matrix power, determinant, rank and matrix inverse. The results of this Chapter are published in [DT08b] and [DT08d].

In Chapter 10.1 we present an application of the quantum walk search schema by Magniez et al. [MNRS07] for finding more than one solution of a search problem. We apply our quantum walk to matrix multiplication, thereby improving a result by Buhrman and Špalek [BŠ06].

In Chapter 10.2 we determine the quantum query complexity of the matrix power resp. matrix power element problem. In the matrix power problem we are given two $n \times n$ matrices A and B and an integer m. One has to decide whether the m'th power of A is the matrix B. In the matrix power element problem we are given a $n \times n$ matrix A and integers i, j, a, m, the task is to decide if the m'th power of A on position (i, j) is the element a. We show that the quantum query complexity of these two problems is $\Theta(n^2)$.

In Chapter 10.3 we present quantum query bounds for matrix inverse, determinant and rank. We use our results from Chapter 10.2 and give a reduction from the matrix power element problem to determinant, matrix inverse and rank problem. Then we get $\Omega(n^2)$ quantum query lower bounds for these problems. Therewith follows that with quantum computation we can not speed up these linear algebra problems in the query model.

10.1 Matrix Multiplication

The quantum walk search algorithm of Theorem 3.3.1 finds only one solution of a search problem. In many practical applications we are interested in more solutions. In this subsection we apply the quantum walk search schema of Theorem 3.3.1 to find more solutions of a search problem in a Johnson graph.

Theorem 10.1.1 *Let $S \subseteq [n]$ be a search problem and let P be a random walk on the Johnson graph $J(n,r)$, where $r = o(n)$. Let M be the class of all r-subsets that contain a solution of S. Then there is a quantum algorithm that finds up to k of the solutions with cost*

$$
\begin{cases}
O\left(s \cdot k + \sqrt{k \cdot \frac{n}{r}}\left(\sqrt{r}u + c\right)\right), & \frac{kr}{n} \leq 1, \\
O\left(s \cdot k + \sqrt{k \cdot \frac{n}{r}}\log n\left(\sqrt{r}u + c\right)\right), & \frac{kr}{n} > 1.
\end{cases}
$$

Proof. Suppose our search problem contains at most k different solutions. We use the quantum walk search schema of Theorem 3.3.1. The result of this quantum search is an element of the marked states M.

The marked state contains a solution x of our search problem. We store this element x in a list. Now we use the quantum walk search schema again for finding another solution. For this task, we modify the oracle in the amplitude amplification, such that a state of the Johnson graph is marked, if it contains a solution which is not yet in the list. We repeat this quantum walk search step $k - 1$ times. The result of this procedure is a list with k different solutions. In case that there are only $l < k$ solutions, possibly $l = 0$, then the algorithm will detect this after the $(l + 1)$-th iteration and output the l solutions found.

Now we compute the quantum cost of this search algorithm. In the $(k - i + 1)$-th iteration of our algorithm, the search problem contains i different solutions. Let M_i be the set of marked states of the Johnson graph with state space X, when there are i different solutions. Let furthermore $\varepsilon_i := \frac{|M_i|}{|X|}$. By Theorem 3.3.1, the cost for finding k solutions is

$$
s \cdot k + \sum_{i=0}^{k-1} \frac{1}{\sqrt{\varepsilon_{k-i}}} \left(\frac{1}{\sqrt{\delta}} u + c \right) = s \cdot k + \Delta_\varepsilon \left(\frac{1}{\sqrt{\delta}} u + c \right)
$$

where

$$
\Delta_\varepsilon := \frac{1}{\sqrt{\varepsilon_1}} + \ldots + \frac{1}{\sqrt{\varepsilon_k}}.
$$

The eigenvalue gap δ of the Johnson graph is $\Theta(1/r)$ for $1 \leq r \leq \frac{n}{2}$, therefore the cost is bounded by

$$
s \cdot k + \Delta_\varepsilon \left(\sqrt{r} u + c \right).
$$

Now we compute the value of Δ_ε. The number of states in the Johnson graph $J(n,r)$ is $|X| = \binom{n}{r}$. If there is one solution, then there are $\binom{n-1}{r-1}$ marked vertices. Suppose there are i solutions of the search problem, then there are

$$|M_i| = \binom{n}{r} - \binom{n-i}{r}$$

marked vertices in the Johnson graph. Then we obtain

$$\varepsilon_i = \frac{|M_i|}{|X|} = 1 - \frac{\binom{n-i}{r}}{\binom{n}{r}} \geq 1 - \left(\frac{n-i}{n}\right)^r .$$

Since $(1 - \frac{x}{n})^n \leq e^{-x}$, we get

$$\varepsilon_i \geq 1 - e^{-\frac{ir}{n}} .$$

Now we get an upper bound for Δ_ε

$$\Delta_\varepsilon \leq \sum_{i=1}^{k} \frac{1}{\sqrt{1 - e^{-\frac{ir}{n}}}} \leq \int_1^{k+1} \frac{1}{\sqrt{1 - e^{-\frac{ir}{n}}}}\, di = \left. \frac{2n}{r}\text{arctanh}\sqrt{1 - e^{-\frac{ir}{n}}} \right|_1^{k+1} .$$

If $\frac{kr}{n} \leq 1$, then $\text{arctanh}\sqrt{1 - e^{-\frac{kr}{n}}} \in O\left(\sqrt{\frac{kr}{n}}\right)$, by using the definition of arctanh and a simple analysis estimation. Therefore it holds

$$\Delta_\varepsilon \leq O\left(\frac{n}{r} \cdot \sqrt{\frac{kr}{n}}\right) = O\left(\sqrt{\frac{kn}{r}}\right) .$$

Otherwise, if $\frac{kr}{n} > 1$, then $\text{arctanh}\sqrt{1 - e^{-\frac{kr}{n}}} \in O\left(\sqrt{\frac{kr}{n}}\log\left(\frac{kr}{n}\right)\right)$, then

$$\Delta_\varepsilon \leq O\left(\sqrt{\frac{kn}{r}}\log n\right) .$$

$\square$

Now we use the idea of the proof of Theorem 10.1.1 and combine it with the bound on the number of marked state in the graph categorical product of two Johnson graphs shown in [BŠ06].

Theorem 10.1.2 *Let $S \subseteq [n] \times [n]$ be a search problem and let P be a random walk on $J(n,r) \times J(n,r)$. Let M be the class of all $(r \times r)$-subsets that contain a solution of S. Then there is a quantum algorithm that finds up to k of the solutions with $r \le n^{2/3}/\min(k, \sqrt{n})^{1/3}$ and cost*

$$
\begin{cases}
O\left(s \cdot k + \frac{n}{r}\sqrt{k}\left(\sqrt{r}u + c\right)\right), & k \le \sqrt{n}, \\
O\left(s \cdot k + \frac{n^{3/4}}{r}k\left(\sqrt{r}u + c\right)\right), & k > \sqrt{n}.
\end{cases}
$$

Proof. The eigenvalue gap δ of $J(n,r) \times J(n,r)$ is $\Theta(1/r)$, for $1 \le r \le \frac{n}{2}$ (see [BŠ06]). Therefore the cost for finds up to k of the solutions is bounded by

$$
s \cdot k + \Delta_\varepsilon \left(\sqrt{r}u + c\right),
$$

where $\Delta_\varepsilon := \frac{1}{\sqrt{\varepsilon_1}} + \ldots + \frac{1}{\sqrt{\varepsilon_k}}$. Now we use the estimation of the value of ε_i by [BŠ06], which holds also for the general search problem in the graph categorical product of two Johnson graphs $J(n,r)$. Therefore

$$
\varepsilon_i = \Omega\left(\frac{r^2}{n^2}q_i\right),
$$

where $q_i \ge \min(i, \sqrt{n})$. Then we get

$$
\Delta_\varepsilon \le \sum_{i=1}^{\sqrt{n}} \frac{n}{r\sqrt{i}} + \sum_{i=\sqrt{n}+1}^{k} \frac{n}{rn^{\frac{1}{4}}} \le
\begin{cases}
\frac{n}{r}\sqrt{k}, & k \le \sqrt{n}, \\
\frac{n^{3/4}}{r}k, & k > \sqrt{n}.
\end{cases}
$$

$\square$

Now we present an application of the Theorem 10.1.2 for matrix multiplication:

Matrix Multiplication: Given two $n \times n$ matrices A and B over any integral domain, compute the product $C = AB$.

The fastest known classical algorithm for computing the product of two matrices works in time $O(n^{2.376})$, see Coppersmith and Winograd [CW90]. Buhrman and Špalek [BŠ06] presented a quantum algorithm which is faster when the number of nonzero elements of the product matrix is $o(n^{0.876})$. The worst case quantum query complexity of their algorithm is $n^{5/3} w \log n$, where w is the number of nonzero entries of the matrix $C = AB$. The expected quantum time complexity of their algorithm is

$$
\begin{cases}
O(n^{5/3} w^{2/3} \log n), & 1 \leq w \leq \sqrt{n} \\
O(n^{3/2} w \log n), & \sqrt{n} \leq w \leq n \\
O(n^2 \sqrt{w} \log n), & n \leq w \leq n^2.
\end{cases}
$$

We present a quantum algorithm that uses Theorem 10.1.2, which improves the worst case quantum query complexity of [BŠ06]. For $1 \leq w \leq n$, the worst case complexity of our algorithm is even better than the expected time complexity of [BŠ06] by a logarithmic factor. We formulate our theorem in terms of the query complexity. By additionally multiplying with random verctors (see [BŠ06]) one can achieve the same time complexity as the query complexity.

Theorem 10.1.3 *There is a quantum algorithm for the matrix multipli-*

cation problem with query complexity of

$$\begin{cases} O(n^{5/3}w^{2/3}), & 1 \le w \le \sqrt{n}, \\ O(n^{3/2}w), & \sqrt{n} < w \le n^2, \end{cases}$$

where w is the number of nonzero entries of the product matrix C.

Proof. Given two $n \times n$ matrices A and B, we want to compute the matrix $C = AB$. At the beginning we set C to the zero-matrix. Our algorithm consists of two main steps. In the first step we search for all wrong entries in the matrix C. In the second step we recompute all wrong entries.

For the first step, we apply Theorem 10.1.2 for finding all nonzero entries in C. Let R, S be two subsets of $[n]$ of size r. We will determine r later. The database is the $r \times r$ matrix $D(R, S) = A|_{R,*} \cdot B|_{*,S}$. The quantum walk takes place on the graph categorical product of two Johnson graphs $J = J(n, r) \times J(n, r)$. The marked vertices of J correspond to pairs (R, S) with $A|_{R,*} \cdot B|_{*,S} \ne C|_{R,S}$. In every step of the walk, we exchange one row and one column of R and S. The setup query cost for the database is $O(rn)$ and the update query cost is $O(n)$. For checking if a vertex is marked, we use Grover search for finding an entry (i, j) with $(A|_{R,*} \cdot B|_{*,S} - C|_{R,S})_{i,j} \ne 0$. Therefore the checking cost is $O(r)$.

Suppose $w \le \sqrt{n}$, then the quantum query complexity of this step is

$$O\left(rnw + \frac{n}{r}\sqrt{w}\left(\sqrt{r}n + r\right)\right).$$

Let $r = \frac{n^{2/3}}{w^{1/3}}$, then we satisfy the condition $r \le n^{2/3}/\min(w, \sqrt{n})^{1/3}$ of Theorem 10.1.2, since $\min(w, \sqrt{n}) \le w$. Since we do not know the number w of nonzero entries, we search in ascending order for $1, 2, 4, \ldots, 2^{\log w - 1}$ nonzero entries. Then the total query complexity for $w \leq \sqrt{n}$ of this iteration is

$$O \left(\sum_{i=0}^{\log w - 1} n^{5/3} \cdot 2^{2i/3} \right) = O(n^{5/3} w^{2/3}).$$

Otherwise, if $w > \sqrt{n}$, then the quantum query complexity is

$$O \left(rnw + \frac{n^{3/4}}{r} w \left(\sqrt{rn} + r \right) \right) = O \left(rnw + \frac{n^{7/4}}{\sqrt{r}} w \right) = O(n^{3/2} w)$$

for $r = \sqrt{n}$. The value of r satisfy the condition of Theorem 10.1.2. In the second step of our algorithm we recompute all wrong entries of C, this can be done in $O(nw)$ queries. □

10.2 Matrix Power

We determine the quantum query complexity of the following two linear algebra problems:

Matrix Power: Given two $n \times n$ matrices A and B and an integer m, decide whether $A^m = B$.

Matrix Power Element: Given a $n \times n$ matrix A and integers i, j, a, m, decide if $(A^m)_{i,j} = a$.

We show that the quantum query complexity of this problem is $\Theta(n^2)$. For this task, we define the following problem for n variables

$x_1, \ldots, x_n \in \{0, 1\}$ and $0 \le a \le n$:

$$\text{EXACT}_n(x_1, \ldots, x_n, a) := \begin{cases} 1, & \text{if } \sum_{i=1}^{n} x_i = a, \\ 0, & \text{otherwise.} \end{cases}$$

Lemma 10.2.1 *The quantum query complexity of* EXACT_n *is* $\Theta(n)$.

Proof. We apply Theorem 4.1.1 to the restriction of EXACT_n to $a = \lfloor n/2 \rfloor$. That is, we consider

$$D = \{ (x_1, \ldots, x_n) \in \{0, 1\}^n \mid \sum_{i=1}^{n} x_i = \lfloor n/2 \rfloor \}.$$

Define $A = D$ and

$$B = \{ (x_1, \ldots, x_n) \in \{0, 1\}^n \mid \sum_{i=1}^{n} x_i = \lfloor n/2 \rfloor + 1 \}.$$

For any sequence $x \in A$, if we change any of the $m = \lfloor n/2 \rfloor$ zeros of x to one, the resulting sequence will be in B. Conversely, for any sequence $x' \in B$, if we change any of the $m' = \lfloor n/2 \rfloor + 1$ ones of x' to zero, the resulting sequence will be in A. Therefore the quantum query complexity of D, and hence of EXACT_n, is $\Omega(\sqrt{m \cdot m'}) = \Omega(n)$. $\qquad \square$

We show in the following how to reduce EXACT_{n^2} to matrix power, in fact, power of 3.

Theorem 10.2.2 *The quantum query complexity of the matrix power and the matrix power element problem is* $\Theta(n^2)$. *This already holds for powers of 3.*

Proof. Given $x_1, \ldots, x_{n^2} \in \{0, 1\}$ and $0 \le a \le n^2$ as input for EXACT_{n^2}, we construct a directed graph G as follows. G has $2n + 2$

nodes. With nodes $1, \ldots, 2n$ we construct a bipartite graph with nodes $1, \ldots, n$ on the left side and nodes $n + 1, \ldots, 2n$ on the right side. For the edges, we consider the variables x_k. Index k can be uniquely written as $k = (i - 1)n + j$, for $1 \le i, j \le n$. Edge $(i, n + j)$ is present in G iff $x_k = 1$. For the remaining two nodes, let $s = 2n + 1$ and $t = 2n + 2$. Add edges from s to all the nodes $1, \ldots, n$ and edges from all nodes $n + 1, \ldots, 2n$ to t. This completes the construction of graph G.

Observe that all paths from s to t in G have length 3 and each such path uniquely corresponds to a variable x_k with value 1. Moreover, there are no further paths of length 3 in G. Let A be the adjacency matrix of G. We conclude that the entry (s, t) of A^3 is the number of paths from s to t in G, and all other entries are 0. Hence we have

$$\text{EXACT}_{n^2}(x_1, \ldots, x_{n^2}, a) = 1 \iff (A^3)_{s,t} = a \iff A^3 = B,$$

where matrix B has (s, t) entry a and 0 elsewhere. $\qquad \square$

10.3 Determinant and Inverse

Next, we consider the determinant, inverse and rank problem:

Determinant: Given a $n \times n$ matrix A, decide whether $\det(A) = 0$.

Inverse: Given a regular $n \times n$ matrix A and integers i, j, a, decide whether $(A^{-1})_{i,j} = a$.

Rank: Given an $n \times n$ matrix A and integer k, decide whether

$\text{rank}(A) = k.$

Theorem 10.3.1 *The quantum query complexity of the determinant and the inverse problem is $\Theta(n^2)$.*

Proof. We slightly modify the standard reduction from matrix power element to the determinant. Let $A = (a_{i,j})$ be a $n \times n$ matrix and a be given. By Theorem 10.2.2, we may assume that A is a 0-1-matrix and we consider the problem whether $(A^3)_{1,n} = a$. We will construct a matrix B such that $(A^3)_{1,n} = \det(B)$.

Interpret A as representing a directed bipartite graph on $2n$ nodes. That is, the nodes are arranged in two columns of n nodes each. In both columns, nodes are numbered from 1 to n. If $a_{i,j} = 1$ then we put an edge from node i in the first column to node j in the second column.

Now, take 3 copies of this graph, put them in a sequence and identify each second column of nodes with the first column of the next graph in the sequence. We have 4 columns of n nodes each so far. Now we add a 5-th column of n nodes as well and connect it by horizontal edges with the 4-th column.

Call the resulting graph G'. Graph G' has $N = 5n$ nodes, and the entry at position $(1, n)$ in A^3 is the number of paths in G' from node 1 in the first column to node n in the last column. Call these two nodes s and t, respectively.

Next, we add an edge from t to s and put self-loops at all nodes except s and t. Call the resulting graph G and let B be the adjacency matrix of G. From combinatorial matrix theory we know that the determinant of B is the signed sum of cycle covers of G. Any cycle cover of G consists of one cycle of length 5 which goes from s to t via the columns in G' and then back to s. The remaining cycles of the cover are self-loops. Therefore each cycle cover corresponds to one path from s to t in G'. The sign of the cycle cover is $(-1)^{N+k}$, where k is the number of cycles in the cover. We have $k = N - 4$. Therefore the sign is 1. We conclude that $\det(B) = \left(A^3\right)_{1,n}$.

Note that the size of B is linear in the size of A. Therefore the lower bound for matrix powering carries over to the determinant.

In order to get a reduction to the matrix inverse problem, we modify graph G from above and add self-loops to nodes s and t. Let H be the resulting graph and $C = (c_{i,j})$ be the adjacency matrix of H. The identity permutation is an additional cycle cover of H compared to G. Therefore we have $\det(C) = \left(A^3\right)_{1,n} + 1$. If C has no inverse, then we have $\det(C) = 0$ and consequently $\left(A^3\right)_{1,n} = -1$. In the following, assume that C has an inverse.

Note that $c_{i,i} = 1$ since all nodes have self-loops. Furthermore, with the convention $s = 1$ and $t = N$ we have $c_{N,1} = 1$ because of the edge from t to s. Note that except for position $(N, 1)$ matrix C is an upper triangular matrix. We consider the Laplace expansion of $\det(C)$. Let $C_{i,j}$ denote the matrix obtained from C by deleting row i and column j.

We consider the expansion for the first column:

$$\det(C) = c_{1,1}\det(C_{1,1}) + (-1)^{N+1}\det(C_{N,1})$$
$$= 1 + (-1)^{N+1}\det(C_{N,1}),$$

because $\det(C_{1,1}) = 1$. Let c be the entry at position $(N,1)$ in C^{-1}. By Cramers rule we have $c = \det(C_{N,1})/\det(C)$. Hence we get

$$\det(C) = 1 + (-1)^{N+1}c \cdot \det(C).$$

We replace $\det(C)$ by $\left(A^3\right)_{1,n} + 1$ and obtain

$$\left(A^3\right)_{1,n}\left((-1)^{N+1} - c\right) = c.$$

It follows that $((-1)^{N+1} - c)$ is non-zero and we have

$$\left(A^3\right)_{1,n} = \frac{c}{(-1)^{N+1} - c}.$$

The size of C is linear in the size of A. Therefore the lower bound for matrix powering carries over to the inverse. □

For $k = n$ we have $\operatorname{rank}(A) = n \Longleftrightarrow \det(A) \neq 0$. Therefore the quantum query complexity of the rank problem follows from the determinant problem.

Corollary 10.3.2 *The quantum query complexity of the rank problem is* $\Theta(n^2)$.

Conclusion

Quantum algorithms have the potential to demonstrate that for some problems quantum computation is more efficient than classical computation. A goal of quantum computing is to determine for which problems quantum computers are faster than classical computers.

In this thesis we studied the quantum query and time complexity of several graph and algebra problems. In the first part of our thesis we presented quantum algorithms for important graph problems. We considered matching problems [Doe08], graph traversal problems [Doe07b, Doe07c] and independent set problems [Doe07a]. Our quantum algorithms improve the best known classical complexity bounds. In particular we improved a maximum matching quantum algorithm by Ambainis and Špalek [AS06].

In the second part of our work we presented quantum complexity bounds for group testing problems [DT07, DT08b]. For a set S and a binary operation on S represented as operation table, we considered the decision problem whether a groupoid, semigroup, monoid or quasigroup is a group. We also proved upper and lower bounds for testing associativity, distributivity and commutativity. In particular, we gave the first application of the new quantum random walk technique by Magniez, Nayak, Roland, and Santha [MNRS07] that improves the previous bounds by Ambainis [Amb04a] and Szegedy [Sze04a]. Furthermore we gave tight quantum query complexity bounds of some important linear algebra problems, like the determinant, rank, matrix inverse and the matrix power problem [DT08a].

In the Appendix we give a summary of the quantum query and the time complexity of the regarded graph and algebra problems. We mention several questions that remain open in this research area:

1. From the $\Omega(n^2)$ lower bound for the determinant it is tempting to conjecture the same bound for the perfect matching and the graph isomorphism problem. Is there a quantum query lower bound of $\Omega(n^2)$ for these problems?

2. Are we able to improve the $O(n^{1.3})$ quantum algorithm for triangle finding?

3. Is there a quantum algorithm for computing blocking flows with running time $O(\sqrt{nm})$? This would give us immediately a better algorithm for the maximum flow problem.

4. Are we able to construct a quantum algorithm which computes a travelling salesman tour in $O(c^n)$ for $c < 2$?

5. Is there a quantum algorithm for the semigroup problem which is better than $O(n^{1.5})$ for $|S'| = n$?

6. Is there a classical or a quantum algorithm for the distributivity problem which is faster than the trivial bounds of $O(n^3)$ resp. $O(n^{1.5})$?

7. Are we able to prove a nontrivial lower bound for the decision problem whether a semigroup or monoid is a group?

8. Can we close the gap between the lower and upper bound of the matrix verification problem.

Appendix: Overview Quantum Complexity

We give a summary of the quantum query (**QQC**) and the time complexity (**QTC**) of the regarded graph and algebra problems.

1. General Problems

Problem	Description	QQC	QTC
Element distinctness [Shi02] [Amb04a]	Given are numbers $x_1, \ldots, x_N$, compute k distinct $i_1, \ldots, i_k$, such that $x_{i_1} = \ldots = x_{i_k}$.	$\Omega(N^{2/3})$ $O(N^{k/(k+1)})$	$O(N^{k/(k+1)} \log^c N)$
Collision [BHT98] [Shi02]	Decide if a function $f : [N] \to [N]$ is one to one or r to one.	$\Theta((N/r)^{1/3})$	$O((N/r)^{1/3} \log N)$
Minimum finding [DH96]	Find the smallest values of $f : [N] \to \mathbb{R}$.	$\Theta(\sqrt{N})$	$O(\sqrt{N} \log N)$
Minimum type finding [DHHM04]	Find d smallest values of $f : [N] \to \mathbb{R}$ with different type.	$\Theta(\sqrt{dN})$	$O(\sqrt{dN} \log N)$
Parity [FGGS98]	Compute the parity of N Boolean values.	$\Theta(N)$	
Sorting [HNS01]	Given are numbers $x_1, \ldots, x_N$, compute $\pi \in S_N$ with $x_{\pi_1}, \ldots, x_{\pi_N}$ is in nondecreasing order.	$\Theta(N \log N)$	
Ordered Searching [HNS01]	Given are numbers $x_1, \ldots, x_N$ in nondecreasing order and $y < x_N$, find the minimal i such that $y \leq x_i$.	$\Theta(\log N)$	

2. Graph Problems

Problem	Description	QQC	QTC
Graph Connectivity [DHHM04]	Decide if G is connected.	**M:** $\Theta(n^{1.5})$ **L:** $\Theta(n)$	**M:** $O(n^{1.5}\log n)$ **L:** $O(n\log n)$
Strong Graph Connectivity [DHHM04]	Decide if G has a directed path between every pair of vertices.	**M:** $\Theta(n^{1.5})$ **L:** $\Omega(\sqrt{nm})$ $O(\sqrt{nm\log n})$	**M:** $O(n^{1.5}\log n)$ **L:** $O(\sqrt{nm}\log^{1.5} n)$
Minimum Spanning Tree [DHHM04]	Compute a minimum spanning tree in G.	**M:** $\Theta(n^{1.5})$ **L:** $\Theta(\sqrt{nm})$	**M:** $O(n^{1.5}\log n)$ **L:** $O(\sqrt{nm}\log n)$
Shortest Paths [DHHM04]	Compute a tree T in G, such that the shortest paths from a vertex v to all the other vertices is in T.	**M:** $\Omega(n^{1.5})$ $O(n^{1.5}\log^2 n)$ **L:** $\Omega(\sqrt{nm})$ $O(\sqrt{nm}\log^2 n)$	**M:** $O(n^{1.5}\log^3 n)$ **L:** $O(\sqrt{nm}\log^3 n)$
Eulerian Graph [Doe07b]	Decide if G has a closed walk that contains every edge of G once.	**M:** $\Theta(n^{1.5})$ **L:** $\Theta(\sqrt{n})$	**M:** $O(n^{1.5}\log n)$ **L:** $O(\sqrt{n}\log n)$
Optimal Postman Tour [Doe07c]	Compute a closed walk of minimum total edge weight that uses each edge at least once.	**M:** $\Omega(n^2)$	**M:** $O(n^{2.5}\log^3 n)$ **L:** $O(n^{1.5}\sqrt{m}\log^3 n)$
Hamiltonian Circuit [BDFLS04] [Doe07b]	Decide if G contains a hamiltonian circuit.	**M:** $\Omega(n^{1.5})$ $O(n^{2n/(n+1)})$	
Traveling Salesman [BDFLS04]	Decide if G has a hamiltionian tour with length k or less.	**M:** $\Omega(n^{1.5})$	
Maximal Matching [Doe08]	Compute a maximal matching in a graph.	**M:** $\Omega(n^{1.5})$ $O(n^{1.5}\log n)$ **L:** $O(\sqrt{nm}\log n)$	**M:** $O(n^{1.5}\log^2 n)$ **L:** $O(\sqrt{nm}\log^2 n)$
Maximum Matching [Doe08]	Compute a maximum matching in a graph.	**M:** $\Omega(n^{1.5})$	**M:** $O(n^2\log^2 n)$ **L:** $O(n\sqrt{m}\log^2 n)$
Max. Weight Bip. Matching [Doe08]	Compute a maximum weight matching in a bipartite graph.	**M:** $\Omega(n^{1.5})$	**M:** $O(n^2 N\log^2 n)$ **L:** $O(n\sqrt{m}N\log^2 n)$

Min. Weight Bip. Matching [Doe08]	Compute a minimum weight perfect matching in a bipartite graph.	**M:** $\Omega(n^{1.5})$	**M:** $O(n^{2.5}\log^3 n)$ **L:** $O(n\sqrt{nm}\log^3 n)$
Triangle finding [BDHHMSW] [MSS05]	Decide if G contains a triangle.	**M:** $\Omega(n)$ $O(n^{1.3})$	**M:** $O(n^{1.5}\log n)$ **L:** $O(\sqrt{nm}\log n)$
Graph Copy [MSS05]	Decide if G contains a copy of a graph H with k vertices.	**M:** $\Omega(n)$ $O(n^{2-2/k})$	
Bipartiteness [Zha04]	Decide if G is bipartite.	**M:** $\Omega(n^{1.5})$	
Maximum Flow [AS06]	Compute a maximum flow in a flow network with integer capacities at most $U \leq n^{1/4}$.	**M:** $\Theta(n^2)$	**M:** $O(n^{\frac{13}{6}}\sqrt[3]{U}\log^2 n)$ **L:** $O(n^{\frac{7}{6}}\sqrt{m}U^{\frac{1}{3}}\log^2 n)$ $O(\sqrt{nU}m\log^2 n)$
Maximal Independent Set [Doe07a]	Compute a maximal independent set in G.	**M:** $\Omega(n^{1.5})$ $O(n^{1.5}\log n)$ **L:** $O(\sqrt{nm}\log n)$	**M:** $O(n^{1.5}\log^2 n)$ **L:** $O(\sqrt{nm}\log^2 n)$
Maximum Independent Set [Doe07a]	Compute a maximum independent set in G.	**M:** $\Omega(n^{1.5})$	**M, L:** $O(1.1488^n)$

3. Algebra Problems

Problem	Description	QQC	QTC
Semigroup I [DT07] [DT08a]	Decide if $S \times S \to S'$ is a semigroup for constant size of S'.	$\Omega(n)$ $O(n^{\frac{5}{4}})$	$O(n^{\frac{3}{2}} \log n)$
Semigroup II [DT07] [DT08a]	Decide if $S \times S \to S'$ is a semigroup.	$\Omega(n)$ $O(n^{\frac{3}{2}})$	$O(n^{\frac{3}{2}} \log n)$
Monoid I [DT07]	Decide if a groupoid is a monoid.	$\Omega(n)$ $O(n^{\frac{3}{2}})$	$O(n^{\frac{3}{2}} \log n)$
Monoid II [DT07]	Decide if a semigroup is a monoid.	$O(n)$	$O(n \log n)$
Quasigroup [DT07] [DT08a]	Decide if a groupoid is a quasigroup.	$\Omega(n)$ $O(n^{\frac{7}{6}})$	$O(n^{\frac{7}{6}} \log n)$
Group I [DT08a]	Decide if a groupoid is a group.	$\Omega(n)$ $O(n \log n)$	$O(n^{\frac{13}{12}} \log^c n)$
Group II [DT07] [DT08a]	Decide if a semigroup/monoid is a group.	$O(n^{\frac{11}{14}} \log n)$	$O(n^{\frac{11}{14}} \log^c n)$
Group III [DT08a]	Decide if a quasigroup/loop is a group.	$\Theta(n)$	$O(n \log n)$
Commut. I [DT08a]	Decide if a groupoid/ semigroup/monoid is commutative.	$\Theta(n)$	$O(n \log n)$
Commut. II [DT08a]	Decide if a quasigroup/group is commutative.	$\widetilde{O}((\log n)^{\frac{2}{3}})$	$\widetilde{O}((\log n)^{\frac{2}{3}})$

4. Linear Algebra Problems

Problem	Description	QQC	QTC
Matrix Verification [BŠ06]	Verify the matrix product $AB = C$.	$\Omega(n^{\frac{3}{2}})$ $O(n^{\frac{5}{3}})$	$O(n^{\frac{5}{3}} \log n)$
Matrix Vector Verification [DT08b]	Verify the matrix-vector product $Ab = c$.	$\Theta(n^{\frac{3}{2}})$	$O(n^{\frac{3}{2}} \log n)$
Matrix Power [DT08b]	Decide if $A^m = B$ for matrices A, B and integer m.	$\Theta(n^2)$	
Matrix Power Element [DT08b]	Decide if $(A^m)_{i,j} = a$ for matrix A and integer i, j, a, m.	$\Theta(n^2)$	
Determinant [DT08b]	Decide if $\det(A) = 0$ for matrix A.	$\Theta(n^2)$	
Inverse [DT08b]	Decide if $(A^{-1})_{i,j} = a$ for matrix A and integers i, j, a.	$\Theta(n^2)$	
Rank [DT08b]	Decide if $\text{rank}(A) = k$ for matrix A and integer k.	$\Theta(n^2)$	

Bibliography

[AA03]   S. Aaronson, A. Ambainis, *Quantum search of spatial structures*, Proceedings of FOCS'03: pages 200-209, 2003.

[AAKV01]   A. Aharonov, A. Ambainis, J. Kempe, U. Vazirani, *Quantum walks on graphs*, Proceedings of STOC'01: pages 50-59, 2001.

[ABNVW01]   A. Ambainis, E. Bach, A. Nayak, A. Vishwanath, J. Watrous, *One dimensional quantum walks*, Proceedings of STOC'01: pages 37-49, 2001.

[ADH97]   L. Adleman, J. DeMarrais, M. Huang, *Quantum computability*, SIAM Journal of Computing 26: pages 1524-1540, 1997.

[AF]   D. Aldous, L. Fill, *Reversible Markov Chains and Random Walks on Graphs*, manuscript, http://www.stat.berkeley.edu/ aldous/RWG/book.html.

[AKR05]   A. Ambainis, J. Kempe, A. Rivosh, *Coins make quantum walks faster*, Proceedings of SODA'05: pages 1099-1108,

2005.

[Amb00]   A. Ambainis, *Quantum query algorithms and lower bounds*, Proceedings of FOTFS III, 2000.

[Amb02]   A. Ambainis, *Quantum Lower Bounds by Quantum Arguments*, Journal of Computer and System Sciences 64: pages 750-767, 2002.

[Amb03]   A. Ambainis, *Quantum walks and their algorithmic applications*, International Journal of Quantum Information 1: pages 507-518, 2003.

[Amb04a]   A. Ambainis, *Quantum walk algorithm for element distinctness*, Proceedings of FOCS'04: pages 22-31, 2004.

[Amb04b]   A. Ambainis, *Quantum Search Algorithms*, SIGACT News, vol. 35: pages 22-35, 2004.

[AS04]   S. Aaronson, Y. Shi, Quantum lower bounds for the collision and the element distinctness problems, Journal of ACM 51: pages 595-605, 2004.

[AS06]   A. Ambainis, R. Špalek, *Quantum Algorithms for Matching and Network Flows*, Proceedings of STACS'06: pages 172-183, 2006.

[BBBV97]   C.H. Bennett, E. Bernstein, G. Brassard, U. Vazirani, *Strengths and weaknesses of quantum computing*, SIAM Journal on Computing 26(5): pages 1510-1523, 1997.

[BBCJPW93] C. Bennett, G. Brassard, C. Crepeau, R. Jozsa, A. Peres, W. Wootters, *Teleporting an unknown quantum state via dual classical and EPR channels*, Physical Review Letters: pages 1895-1899, 1993.

[BBCMW01] R. Beals, H. Buhrman, R. Cleve, M. Mosca, R. de Wolf, *Quantum lower bounds by polynomials*, Journal of ACM 48: pages 778-797, 2001.

[BBHT98] M. Boyer, G. Brassard, P. Høyer, A. Tapp, *Tight bounds on quantum searching*, Fortschritte Der Physik 46(4-5): pages 493-505, 1998.

[BCWZ99] H. Buhrman, R. Cleve, R. de Wolf, Ch. Zalka, *Bounds for Small-Error and Zero-Error Quantum Algorithms*, Proceedings of FOCS'99: pages 358-368, 1999.

[BDFLS04] A. Berzina, A. Dubrovsky, R. Freivalds, L. Lace, O. Scegulnaja, *Quantum Query Complexity for Some Graph Problems*, Proceedings of SOFSEM'04: pages 140-150, 2004.

[BDHHMSW01] H. Buhrman, C. Dürr, M Heiligman, P. Høyer, F. Magniez, M. Santha, R. de Wolf, *Quantum Algorithms for Element Distinctness*, Proceedings of CCC'01: pages 131-137, 2001.

[Bei99]   R. Beigel, *Finding maximum independent sets in sparse and general graphs*, Proceedings of SODA'99: pages 856-857, 1999.

[Beh00]   E. Behrends, *Introduction to Markov Chains*, Vieweg, Braunschweig, 2000.

[Ben80]   P. Benioff, *The Computer as a Physical System: A Microscopic Quantum Mechanical Hamiltonian Model of Computers as Represented by Turing Maschines*, Journal of Statistical Physics 22: pages 563-591, 1980.

[BHMT02]   G. Brassard, P. Hóyer, M. Mosca, A. Tapp, *Quantum amplitude amplification and estimation*, AMS Contemporary Mathematics, Vol. 305: pages 53-74, 2002.

[BHT98]   G. Brassard, P. Høyer, A. Tapp, *Quantum cryptanalysis of hash and claw-free functions*, Proceedings of LATIN'98: pages 163-169, 1998.

[BM76]   J. Bondy, U. Murty, *Graph Theory with Applications*, North-Holland, New York, 1976.

[BŠ06]   H. Buhrman, R. Špalek, *Quantum Verification of Matrix Products*, Proceedings of SODA'06: pages 880-889, 2006.

[BV97]   E. Bernstein, U. Vazirani, *Quantum complexity theory*, SIAM Journal on Computing 26: pages 1411-1473, 1997.

[BW02]   H. Buhrman, R. de Wolf, *Complexity measures and decision tree complexity: a survey*, Theoretical Computer Science 288: pages 21-43, 2002.

[CCDFGS02] A.M. Childs, R. Cleve, E. Deotto, E. Farhi, S. Gutmann, D.A. Spielman, *Exponential algorithmic speedup by quantum walk*, Proceedings of STOC'03: pages 59-68, 2003.

[CCPS98]   W.J. Cook, W.H. Cunningham, W.R. Pulleyblank, A. Schrijver, *Combinatorial Optimization*, John Wiley & SONS, INC, New York, 1998.

[CE03]   A. Childs, J. Eisenberg, *Quantum algorithms for subset finding*, Technical Report arXiv:quant-ph/0311038, 2003.

[CEMM98]   R. Cleve, A. Ekert, C. Macchiavello, M. Mosca, *Quantum algorithms revisited*, Proceedings of the Royal Society A: Mathematical, Physical and Engineering Sciences, 454: pages 339-354, 1998.

[CG03]   A. Childs, J. Goldstone, *Spatial search by quantum walk*, Technical Report arXiv:quant-ph/0306054, 2003.

[CM97]   C. Moore, J. P. Crutchfield, *Quantum Automata and Quantum Grammars*, Theoretical Computer Science, Santa Fe Institute Working Paper, 1997.

[CP61]   A.H. Clifford, G.B. Preston, *The Algebraic Theory of Semigroups*, American Mathematical Society, 1961.

197

[CVZLL98]   L. Chuang, L. M. K. Vandersypen, X. L. Zhou, D. W. Leung, S. Lloyd, *Experimental realization of a quantum algorithm*, Nature, 393 no. 6681: pages. 143-146, 1998.

[CW90]   D. Coppersmith, S. Winograd, *Matrix multiplication via arithmetic progressions*, Journal of symbolic computation 9: pages 251-280, 1990.

[Deu85]   D. Deutsch, *Quantum theory, the Church-Turing principle and the universal quantum computer*, Proc. R. Soc. Lond. A, 400: pages 97-117, 1985.

[DH96]   C. Dürr, P. Høyer, *A quantum algorithm for finding the minimum*, Technical Report arXiv:quant-ph/9607014, 1996.

[DHHM04]   C. Dürr, M. Heiligman, P. Høyer, M. Mhalla, *Quantum query complexity of some graph problems*, Proceedings of ICALP'04: pages 481-493, 2004.

[DHTW08]   S. Dörn, D. Haase, J. Torán, F. Wagner, *Isomorphism and Factorization-Classical and Quantum Algorithms*, In Mathematical Analysis of Evolution, Information and Complexity, Wiley, 2008.

[Die00]   R. Diestel, *Graph Theory*, Electronic Edition, 2000.

[DJ92]   D. Deutsch, R. Jozsa, *Rapid Solution of Problems by Quantum Computations*, Proc. R. Soc. London A, 492: pages 553-558, 1992.

[Doe07a]   S. Dörn, *Quantum Complexity Bounds of Independent Set Problems*, Proceedings of SOFSEM'07 (SRF): pages 25-36, 2007.

[Doe07b]   S. Dörn, *Quantum Algorithms for Graph Traversals and Related Problems*, Proceedings of CIE'07: pages 123-131, 2007.

[Doe07c]   S. Dörn, *Quantum Algorithms for Optimal Graph Traversal Problems*, Proceedings of Quantum Information and Computation V, 2007.

[DT07]   S. Dörn, T. Thierauf, *The Quantum Query Complexity of Algebraic Properties*, Proceedings of FCT'07: pages 250-260, 2007.

[Doe08]   S. Dörn, *Quantum Algorithms for Matching Problems*, to appear at Theory of Computing Systems, 2008.

[DT08a]   S. Dörn, T. Thierauf, *The Quantum Complexity of Group Testing*, Proceedings of SOFSEM'08: pages 506-518, 2008.

[DT08b]   S. Dörn, T. Thierauf, *On the Quantum Query Complexity of Determinant*, submitted to journal, 2008.

[DT08c]   S. Dörn, T. Thierauf, *Quantum Algorithms for Algebraic Problems*, submitted to journal, 2008.

[DT08d]   S. Dörn, T. Thierauf, *Search for k Elements via Quantum Walk* Preprint, 2008.

[EJ73]   J. Edmonds and E.L. Johnson, *Matching Euler tours and the Chinese postman*, Math. Programming 5: pages 88-124, 1973.

[Epp03]   D. Eppstein, *The traveling salesman problem for cubic graphs*, Lecture Notes in Computer Science 2748: pages 307-318, Springer, 2003.

[ET75]   S. Even, R.E. Tarjan, *Network flow and testing graph connectivity*, SIAM Journal on Computing 4: pages 507-518, 1975.

[Fey82]   R. Feynman, *Simulating physics with computers*, International Journal Theoretical Physics 21: pages 467-488, 1982.

[FG98]   E. Farhi, S. Gutmann, *Quantum computation and decision trees*, Physical Review A 58: pages 915-928, 1998.

[FGK06]   F. Fomin, F. Grandoni, D. Kratsch, *Measure and conquer: a simple $O(2^{0.288n})$ independent set algorithm*, Proceedings of SODA'06: pages 18-25, 2006.

[FR98]   L. Fortnow, J. D. Rogers, *Complexity limitations on quantum computation*, Journal of Computer and System Sciences, 59(2): pages 240-252, 1999.

[Gab90]   H. N. Gabow, *Data structures for weighted matching and nearest common ancestors with linking*, Proceedings of the 1st Annual ACM-SIAM Symposium on Discrete Algorithms: pages 434-443, 1990.

[GJ79]   M. R. Garey, D. S. Johnson, *Computers and Intractability*, Bell Telephone Laboratories, 1979.

[Gro96]   L. Grover, *A fast mechanical algorithm for database search*, Proceedings of STOC'96: pages 212-219, 1996.

[Gro02]   L. Grover, *Tradeoffs in the quantum search algorithm*, Technical Report, arXiv:quant-ph/0201152, 2002.

[Gru99]   J. Gruska, *Quantum Computing*, The McGraw-Hill Companies, London, 1999.

[GT89]   H.N. Gabow, R.E. Tarjan, *Faster scaling algorithms for network problems*, SIAM Journal on Computing 18: pages 1013-1036, 1989.

[Gua62]   M. Guan, *Graphic programming using odd and even points*, Chinese Math. 1: pages 273-277, 1962.

[GY99]   J. Gross, J. Yellen, *Graph Theory and its Applications*, CRC Press, London 1999.

[Haj91]   P. Hajnal, *An $n^{4/3}$ lower bound on the randomized complexity of graph properties*, Combinatorica 11: pages 131-143, 1991.

[HK62]   M. Held, R.M. Karp, *A dynamic programming approach to sequencing problems* Journal of SIAM 10: pages 196-210, 1962.

[HK73]   J. E. Hopcroft, R. M. Karp, *An $n^{5/2}$ algorithm for maximum matchings in bipartite graphs*, SIAM Journal on Computing 2(4): pages 225-231, 1973.

[HMW03]   P. Høyer, M. Mosca, R. de Wolf, *Quantum search on bounded-error inputs*, Proceedings of ICALP'03: pages 291-299, 2003.

[HNS01]   P. Høyer, J. Neerbek, Y. Shi, *Quantum complexities of ordered searching, sorting, and element distinctness*, Proceedings of ICALP'01: pages 346-357, 2001.

[Hom05]   M. Homeister, *Quantum Computing verstehen*, Vieweg, 2005.

[HS01]   Y. Hardy, W-H. Steeb, *Classical and Quantum Computing*, Birkhäuser, Basel, 2001.

[HŠ05]   P. Høyer and R. Špalek, *Lower bounds on quantum query complexity*, Bulletin of the European Association for Theoretical Computer Science 87: pages 78-103, 2005.

[Jia86]   T. Jian, *An $O(2^{0.304n})$ algorithm for solving maximum independent set problem*, IEEE Transactions on Computers 35: pages 847-851, 1986.

[Kem03a]   J. Kempe, *Quantum random walks - an introductory overview*, Contemporary Physics 44(4): pages 307-327, 2003.

[Kem03b]   J. Kempe, *Quantum random walks hit exponentially faster*, Lecture Notes in Computer Science 2764: pages 354-369, 2003.

[Kin88]   V. King, *Lower bounds on the complexity of graph properties*, Proceedings of STOC'88: pages 468-476, 1988.

[KLST01]   M.-Y. Kao, T.-W. Lam, W.-K. Sung, H.-F. Ting, *A Decomposition Theorem for Maximum Weight Bipartite Matchings*, SIAM Journal on Computing 31: pages 18-26, 2001.

[Knu03]   D.E. Knuth, *Combinatorial matrices*, Selected Papers on Discrete Mathematics, volume 106 of CSLI Lecture Notes, Stanford University, 2003.

[KST93]   J. Köbler, U. Schöning, J. Torán, *The Graph Isomorphism Problem - Its Structural Complexity*, Birkhäuser, 1993.

[KSW03]   J. Kempe, N. Shenvi, K.B. Whaley, *Quantum Random-Walk Search Algorithm*, Physical Review Letters A, Vol. 67 (5), 2003.

[Kut05]   S. Kutin, *Quantum Lower Bound for the Collision Problem with Small Range*, Theory of Computing, Volumne 1: pages 29-36, 2005.

[LY94]   L. Lovász and N. Young, *Lecture notes on evasiveness of graph properties*, Technical report, Princeton University, 2002.

[MM05]   D.C. Marinescu, G.M. Marinescu, *Approaching Quantum Computing*, Pearson, New Jersey, 2005.

[MN05]   F. Magniez, A. Nayak, *Quantum complexity of testing group commutativity*, Proceedings of ICALP'05: pages 1312-1324, 2005.

[MNRS07]   F. Magniez, A. Nayak, J. Roland, M. Santha, *Search via Quantum Walk*, Proceedings of STOC'07: pages: 575-584, 2007.

[MR95]   R. Motwani, P. Raghavan, *Randomized Algorithms*, Cambridge University Press, 1995.

[MS04]   M. Mucha, P. Sankowski, *Maximum matchings via Gaussian elimination*, Proceedings of FOCS'04: pages 248-255, 2004.

[MSS05]   F. Magniez, M. Santha, M. Szegedy, *Quantum Algorithms for the Triangle Problem*, Proceedings of SODA'05: pages 1109-1117, 2005.

[MV80]   S. Micali, V.V. Vazirani, *An $O(\sqrt{n}m)$ Algorithm for Finding Maximum Matching in General Graphs*, Proceedings of FOCS'80: pages 17-27, 1980.

[NC03]   M.A. Nielsen, I. L. Chuang, *Quantum Computation and Quantum Information*, Cambridge University Press, 2003.

[Pap76]   C.H. Papadimitriou, *On the complexity of edge traversing*, JACM 23: pages 544-554, 1973.

[Rob01]   J.M. Robson, *Finding a maximum independent set in time $O(2^{n/4})$*, Manuscript, 2001.

[RS05]   V. Raman, S. Saurabh, *Efficient Exponential Algorithms Through Enumeration of Maximal Independent Sets*, Technical Report, 2005.

[Shi02]   Y. Shi, *Quantum lower bounds for the collision and the element distinctness problems*, Proceedings of FOCS'02: pages 513-519, 2002.

[Sho94]   P. Shor, *Algorithms for quantum computation: discrete logarithms and factoring*, Proceedings of FOCS'94: pages 124-134, 1994.

[Sho97]   P. Shor, *Polynomial-Time Algorithms for Prime Factor-*
            ization and Discrete Logarithms on a Quantum Computer,
            SIAM Journal on Computing 26: pages 1484-1509, 1997.

[Sin93]   A. Sinclair, *Algorithms for Random Generation and*
            Counting: A Markov Chain Approach, Birkhäuser,
            Boston, 1993.

[Špa06]   R. Špalek, *Quantum Algorithms, Lower Bounds, and*
            Time-Space Tradeoffs, Ph.D These, 2006.

[ŠS06]   R. Špalek, M. Szegedy, *All quantum adversary methods are*
            equivalent, Theory of Computing 2(1): pages 1-18, 2006.

[SYZ04]   X. Sun, A.C. Yao, S. Zhang, *Graph properties and circular*
            functions: How low can quantum query complexity go?,
            Proceedings of CCC'04: pages 286-293, 2004.

[Sze04a]   M. Szegedy, *Quantum speed-up of Markov chain based al-*
            gorithms, Proceedings of FOCS'04: pages 32-41, 2004.

[Sze04b]   M. Szegedy, *Spectra of Quantized Walk and a $\sqrt{\delta\varepsilon}$ rule*,
            Technical Report arXiv:quant-ph/0401053, 2004.

[RS00]   S. Rajagopalan, L. J. Schulman, *Verification of identities*,
            SIAM J. Computing 29(4): pages 1155-1163, 2000.

[TT77]   R.E. Tarjan, A.E. Trojanowski, *Finding a maximum inde-*
            pendent set, SIAM Journal on Computing 6: pages 537-
            546, 1977.

[Wat01]   J. Watrous, *Quantum algorithms for solvable groups*, Proceedings of STOC'01: pages 60-67, 2001.

[Weg03]   I. Wegner, *Komplexitätstheorie*, Springer, Berlin, 2003.

[Woe03]   G.J Woeginger, *Exact algorithms for NP-hard problems: A survey*, Combinatorial Optimization: pages 185-207, Springer, 2003.

[Yao93]   A. Yao, *Quantum circuit complexity*, Proceedings of FOCS'03: pages 352-361, 1993.

[Zha04]   S. Zhang, *On the power of Ambainis's lower bounds*, Proceedings of ICALP'04, Lecture Notes in Computer Science 3142: pages 1238-1250, 2004.

Glossary of Notation

$\lvert \cdot \rvert$	Number of elements in a set
$\{\cdot,\cdot\}$	Undirected edge
$(\cdot,\cdot)$	Directed edge
$\lVert \cdot \rVert$	l_2-norm
$\backslash$	Set difference
$\emptyset$	Empty set
$\oplus$	Direct sum
$\otimes$	Tensor product
$\lvert\cdot\rangle$	Ket-vector
$\langle\cdot\rvert$	Bra-vector
$[n]$	$= \{1,2,\ldots,n\}$
$\mathrm{Prob}(\cdot)$	Probability
$O(\cdot)$	Complexity measure
$\Omega(\cdot)$	Complexity measure
$\Theta(\cdot)$	Complexity measure
$\mathbb{N},\mathbb{Z},\mathbb{R},\mathbb{C}$	Set of natural, integers, real and complex numbers
^{T}	Transpose of matrix
†	Complex conjugate of matrix

$\mathcal{A}$	Quantum algorithm
α	Rotation angle Grover iteration
c	Checking cost quantum walk
$c(G)$	Number of components in graph G
$\chi(A)(a)$	$= 1$, if $a \in A$; $= 0$ otherwise
$d_G(v)$	Degree of vertex v in G
$d_G^+(v)$	In-degree of vertex v in G
$d_G^-(v)$	Out-degree of vertex v in G
D	Database quantum walk search scheme
D_N	Diffusion operator of size $N \times N$
$D(f), R_0(f), R_2(f)$	Exact, zero-error, bounded error query complexity
$Q(f), Q_0(f), Q_2(f)$	Exact, zero-error, bounded error quantum query complexity
$D(P)$	Discriminate of Markov chain P
δ	Spectral gap
$\delta_{x,y}$	$= 1$, if $x = y$; $= 0$ otherwise
ε	Probability of a marked state
$E(G)$	Edge set of graph G
f_P	Search function
G	directed or undirected graph
G_{-e}	Deleting edge e of G
G_{-v}	Deleting vertex v of G

G_{-S}	Deleting vertex/edge set S of G
$G[\cdot]$	Vertex/edge induced subgraph of G
H, H_n	Hadamard matrix with n Qubits
$\mathcal{H}$	Hilbert space
$\mathcal{H}_2$	Two dimensional space $\mathcal{C}_2$
I, I_N	Identity matrix of size $N \times N$
k	Number of solution in a search problem
$J(n,r)$	Johnson graph
M	Set of marked states in Johnson graph
N	Search space
P	Search problem
$P_{i,j}$	Transition matrix of Markov chain
π	Stationary distribution
Π	Orthogonal projector
Q	Amplitude amplification operator
ref	Reflection operator
s	Setup cost quantum walk
$s_x(f), \bar{s}(f)$	sensitivity, average sensitivity
$(S, \circ)$	Groupoid
u	Update cost quantum walk
U_f, U_0	Phase flip or oracle transformation
$V, V(G)$	Vertex set of graph G
W	Quantum walk operator

x^*	Solution of search problem
X	State space of Markov chain

Index

Amplitude amplification, 51

Blocking flow, 92
Breadth first search, 79

Claw finding, 53
Commutativity problem, 164
Computation
   bounded-error, 34
   exactly, 34
   zero-error, 34

Depth first search, 78
Distributivity problem, 165

Element distinctness, 53, 59
Entanglement, 10, 27
EPR-pair, 27
Eulerian graph, 119

Graph
   copy, 61
   notation, 20

property, 75
   monotone, 75
Graph connectivity, 81
Grover
   algorithm, 39
   diffusion, 39
   iteration, 39
   operator, 40
   search complexity, 46

Hadamard matrix, 28
Hamiltonian circuit, 124
Hilbert space
   direct sum, 26
   tensor power, 26
   tensor product, 26

Identity problem, 150
Independent set
   maximal, 134
   maximum, 138

Markov chain, 23
   aperiodic, 23
   ergodic, 24
   irreducible, 23
   reversible, 61
Matching
   maximal, 99
   maximum, 99
   maximum weight bipartite, 109
   minimum weight perfect bipartite, 109
Matrix
   determinant, 178
   inverse, 178
   multiplication, 174
   power, 176
   power element, 176
   Rank, 178
   verification, 60
Maximum flow, 91
Measurement
   computional basis, 29
   POVM, 31
   projective, 30

   projector, 30
Minima finding, 48
Minimum spanning tree, 87
Minimum weight cover, 110
Monoid problem, 151

Operator
   density, 31
   linear, 28
   unitary, 28
Optimal postman tour, 120

Phase
   flip, 39
Polynomial method, 72
Product state, 27
Project scheduling, 131

Quantum
   parallelism, 9
   query complexity, 33
   time complexity, 33
Quantum system, 26
Quantum walk
   search scheme, 57
   database, 56

discriminant, 62

finding all solutions, 170

Qubit

amplitudes, 25

computational basis, 25

state, 25

state space, 25

Semigroup problem, 145

Shortest paths, 88

Strong connectivity, 85

Topological numbering, 80

Travelling salesman problem, 126

Triangle finding, 54, 60

Vector

bra, 26

ket, 26

www.ingramcontent.com/pod-product-compliance
Lightning Source LLC
Chambersburg PA
CBHW071425050326
40689CB00010B/1983